Computer Communications and Networks

Series editor

A.J. Sammes
Centre for Forensic Computing
Cranfield University, Shrivenham campus
Swindon, UK

The **Computer Communications and Networks** series is a range of textbooks, monographs and handbooks. It sets out to provide students, researchers, and non-specialists alike with a sure grounding in current knowledge, together with comprehensible access to the latest developments in computer communications and networking.

Emphasis is placed on clear and explanatory styles that support a tutorial approach, so that even the most complex of topics is presented in a lucid and intelligible manner.

More information about this series at http://www.springer.com/series/4198

K.G. Srinivasa · Anil Kumar Muppalla

Guide to High Performance Distributed Computing

Case Studies with Hadoop, Scalding and Spark

 Springer

K.G. Srinivasa
M.S. Ramaiah Institute of Technology
Bangalore
India

Anil Kumar Muppalla
M.S. Ramaiah Institute of Technology
Bangalore
India

ISSN 1617-7975   ISSN 2197-8433   (electronic)
Computer Communications and Networks
ISBN 978-3-319-38347-7   ISBN 978-3-319-13497-0   (eBook)
DOI 10.1007/978-3-319-13497-0

Springer International Publishing AG Switzerland is part of Springer Science+Business Media (www.springer.com)

Dedicated to Oneness

Preface

Overview

As the use of computers became widespread in the last twenty years, there has been an avalanche of digital data generated. The advent of digitization of all equipments and tools in homes and industry have also contributed to the growth of digital data. The demand to store, process and analyze this huge, growing data is answered by a host of tools in the market. On the hardware front the High Performance Computing (HPC) systems that function above tera-floating-point operations per second undertake the task of managing huge data. HPC systems needs to work in distributed environment as single machine cannot handle the complex nature of its operations. There are two trends in achieving the teraflop scale operations in a distributed way. Connecting computers via global network and handling the complex task of data management in distributed way is one approach. In other approach dedicated processors are kept close to each other thereby saving the data transfer time between the machines. The convergence of both trends is fast emerging and promises to provide faster, efficient hardware solutions to the problems of handling voluminous data.

The popular software solution to the problem of huge data management has been Apache Hadoop. Hadoop's ecosystem consists of Hadoop Distributed File System (HDFS), MapReduce framework with support for multiple data formats and data sources, unit testing, clustering variants and related projects like Pig, Hive etc. It provides tools for life-cycle management of data including storage and processing. The strength of Hadoop is that it is built to manage very large amounts of data through a distributed model. It can also work with unstructured data which makes it attractive. Combined with a HPC backbone, Hadoop can make the task of handling huge data very easy.

Today there are many high level Hadoop frameworks like Pig, Hive, Scoobi, Scrunch, Cascalog, Scalding and Spark that which make it easy to use Hadoop. Most of them are supported by well known organizations like Yahoo (Pig), Facebook (Hive), Cloudera (Scrunch) and Twitter (Scalding) demonstrating the wide patronage Hadoop enjoys in the industry. These frameworks use the basic Hadoop modules like HDFS and MapReduce but provides an easy method to manage complex data processing jobs by creating an abstraction to hide the complexities of Hadoop modules. An example of such abstraction is Cascading. Many specific languages are built using the framework of Cascading. One such implementation by Twitter is called Scalding which it uses to query large data set like tweets stored in HDFS.

Data storage in Hadoop and Scalding is mostly disk based. This architecture impacts the performance due to long seek/transfer time of data. If data is read from disk and then held in memory where they can also be cached, the performance of the system will increase manifold. Spark implements this concept and claims it is 100x faster than MapReduce in memory and 10x faster on disk. Spark uses the basic abstraction of Resilient Distributed Datasets which are distributed immutable collections. Since Spark stores data in memory iterative algorithms in data mining and machine learning can be performed efficiently.

Objectives

The aim of this book is to present the required skills to set up and build large scale distributed processing systems using the free and open source tools and technologies like Hadoop, Scalding, Spark. The key objectives for this book include:

- Capturing the state of the art in building high performance distributed computing systems using Hadoop, Scalding and Spark

- Providing relevant theoretical software frameworks and practical approaches

- Providing guidance and best practices for students and practitioners of free and open source software technologies like Hadoop, Scalding and Spark

- Advancing the understanding of building scalable software systems for large scale data processing as relevant to the emerging new paradigm of High Performance Distributed Computing (HPDC)

Organization

There are 8 chapters in A Guide To High Performance Distributed Computing Case Studies with Hadoop, Scalding and Spark. These are organized in two parts.

Part I: Programming fundamentals of High Performance Distributed Computing

Chapter 1 covers the basics of distributed systems which form the backbone of modern HPDC paradigms like Cloud Computing, Grid/Cluster Systems. It starts by discussing various forms of distributed systems and explaining their generic architecture. Distributed file systems which form the central theme of such design are also covered. The technical challenges encountered in their development and the recent trends in this domain are also dealt with a host of relevant examples.

The discussion on the overview of Hadoop ecosystem in Chapter 2 is followed by a step-by-step instruction on its installation, programming and execution. Chapter 3 starts by describing the core of Spark which is Resilient Distributed Databases. The installation, programming API and some examples are also covered in this chapter. Hadoop streaming is the focus of Chapter 4 which also covers working with Scalding. Using Python with Hadoop and Spark is also discussed.

Part II: Case studies using Hadoop, Scalding and Spark

That the current book does not limit itself to explaining the basic theoretical foundations and presenting sample programs is its biggest advantage. There are four case studies presented in this book which covers a host of application domains and computational approaches so as to convert any doubter into a believer of Scalding and Spark. Chapter 5 takes up the task of implementing K-Means Clustering Algorithm while Chapter 6 covers data classification problems using Naive-Bayes classifier. Continuing the coverage of data mining and machine learning approaches in distributed systems using Scalding and Spark, regression analysis is covered in Chapter 7.

Recommender systems have become very popular today in various domains. They automate the task of middleman who can connect two otherwise disjoint entities. This is becoming much needed feature in all modern networked applications in shopping, searching and publishing. A working recommender system should not only have a strong computational engine but should also be scalable at real-time. Chapter 8 explains the process of creating such a recommender system using Scalding and Spark.

Target Audience

A Guide To High Performance Distributed Computing Case Studies with Hadoop, Scalding and Spark has been developed to support a number of potential audiences, including the following:

- Software Engineers and Application Developers

- Students and University Lecturers

- Contributors to Free and Open Source Software

- Researchers

Code Repository

The complete list of source code and datasets used in this book can be found here
`https://github.com/4nil/hpdc-scalding-spark`

Bangalore, India   *Srinivasa K G*
September 2014   *Anil Kumar Muppalla*

About the Authors

K G Srinivasa

Srinivasa K G received his PhD in Computer Science and Engineering from Bangalore University in 2007. He is now working as a Professor and Head in the Department of Computer Science and Engineering, M S Ramaiah Institute of Technology, Bangalore. He is the recipient of All India Council for Technical Education - Career Award for Young Teachers, Indian Society of Technical Education ISGITS National Award for Best Research Work Done by Young Teachers, Institution of Engineers(India) IEI Young Engineer Award in Computer Engineering, Rajarambapu Patil National Award for Promising Engineering Teacher Award from ISTE - 2012, IMS Singapore Visiting Scientist Fellowship Award. He has published more than hundred research papers in International Conferences and Journals. He has visited many Universities abroad as a visiting researcher He has visited University of Oklahoma, USA, Iowa State University, USA, Hong Kong University, Korean University, National University of Singapore are few prominent visits. He has authored two books namely File Structures using C++ by TMH and Soft Computer for Data Mining Applications LNAI Series Springer. He has been awarded BOYSCAST Fellowship by DST, for conducting collaborative Research with Clouds Laboratory in University of Melbourne in the area of Cloud Computing. He is the principal Investigator for many funded projects from UGC, DRDO, and DST. His research areas include Data Mining, Machine Learning, High Performance Computing and Cloud Computing. He is the Senior Member of IEEE and ACM. He can be reached at *kgsrinivas@msrit.edu*

Anil Kumar Muppalla

Mr. Anil Muppalla is a researcher and author. He holds degree in Computer Science and Engineering. He is a developer and software consultant for many industries. He is also active researcher and published many papers in international conferences and journals. His skills include application development using Hadoop, Scalding and Spark. He can be contacted at *anil@msrit.edu*.

Acknowledgements

The authors acknowledge the help and support of the following colleagues during the preparation of this book:

- Shri. M. R. Seetharam, Director, M S Ramaiah Institute of Technology

- Shri. M. R. Ramaiah, Director, M S Ramaiah Institute of Technology

- Shri. S. M. Acharya, Chief Executive, M S Ramaiah Institute of Technology

- Dr S. Y. Kulkarni, Principal, M S Ramaiah Institute of Technology

- Dr NVR Naidu, Vice-Principal, M S Ramaiah Institute of Technology

- Dr T. V. Suresh Kumar, Registrar, M S Ramaiah Institute of Technology

We thank all the faculty of Department of CSE, MSRIT for their inspiration and encouragement during the preparation of this book. We are grateful to Mr P M Krishnaraj and Dr Siddesh G. M. for their guidance in development of this book. We would also like to thank Mr Nikhil and Mr Maaz for their timely support in composing this book. We are indebted to the Scalding and Spark community for the continuous support during the development of this book.

Grateful thanks are also due to our family members for their support and understanding.

Contents

Part I
Programming Fundamentals of High Performance Distributed Computing

Chapter 1

Introduction

Distributed Computing focuses on a range of ideas and topics. This chapter identifies several properties of distributed systems in general, it also discusses briefly the different types of systems shedding light on popular architectures used in successful distributed system arrangements. It further goes on to identify several challenges and hints at several research areas. The chapters ends with trends and examples where distributed systems have perhaps contributed immensely.

High-Performance Distributed Computing (HPDC) refers to any computational activity that requires many computers to execute a task. The major applications of HPDC include Data Storage and Analysis, Data Mining, Simulation and Modeling, Scientific Calculations, Bio-informatics, Big Data Challenges, Complex Visualizations, and many more.

Early High-performance computing (HPC) systems were more concerned with executing code on parallel architectures aligned with distributed systems. Now the focus has shifted to enabling transparent and efficient utilization of infrastructural capabilities of distributed computing architectures like clusters/grids/clouds.

As hardcoding HPC applications using traditional programming languages have become less favorable, the emergence of distributed software frameworks like Hadoop and Spark help in productive application development on large-scale HPC systems. Functional programming models like MapReduce can easily be implemented on HPC clusters using Hadoop and Spark. The development of these frameworks largely derives inspiration from distributed computing principles.

© Springer International Publishing Switzerland 2015
K.G. Srinivasa and A.K. Muppalla, *Guide to High Performance Distributed Computing*, Computer Communications and Networks, DOI 10.1007/978-3-319-13497-0_1

1.1 Distributed Systems

Distributed computing is a branch of computer science that studies several aspects of computing in distributed system arrangements. It is an arrangement of computers that are networked such that communication between the nodes (computers) is due to complex message passing interfaces. A distributed system is chosen typically to handle datasets and problems that require several hundred computers working in unison, achieving a common goal. A system like this offers a wide range of challenges and problems. Distributed Computing has taken its rightful place as one of the coveted research areas.

While this is an attempt at defining a distributed system, it is certainly not complete. While debating on the physical aspects of a distributed system by calling a system distributed if the system I/O is far from the processor or if the memory is present at a distant geographical location from the processor. In either case, distribution of physical components alone does not fully qualify as criterion for the definition. To alleviate this it is now acceptable to characterize a distributed system using its logical and functional capabilities. These capabilities are usually based on the following set of criteria but are certainly not limited to them:

1. **Multiple Processes:** The system consists of more than one sequential process. These processes can be either system or user defined, but each process should have an independent thread of control, either explicit or implicit.

2. **Interprocess Communication:** Communication channels between processes are vital for a distributed system. The channel's reliability and the time delay between messages depends on the physical characteristics of the links either on a single node or over a networked arrangement [1]. Generally this involves mechanisms on: a) *kernel space* - shared memory, message passing, semaphores, etc and b) *user space* - daemons for network transfer, distributed shared memory, etc.

3. **Disjoint Address Spaces:** Processes may have disjoint address spaces. Shared-memory multiprocessors does alone does not qualify as the true representation of a distributed computing system, although shared memory can be implemented using messages [2].

Although the discussion so far has presented valid criteria for distributed system characterization it is certainly not adequate. Some items like managing system-wide interprocess communication, data or network security are also important for distributed arrangement. Process runtime management and control of user-defined computations are also important while characterizing distributed systems. Nevertheless, these ideas are sufficient for a computation to be logically distributed. Physical distribution of the systems are only a prerequisite for logical distribution.

Computers and systems of all kinds are networked. A good example of network is the Internet to which many of the smaller networks like mobile phone networks,

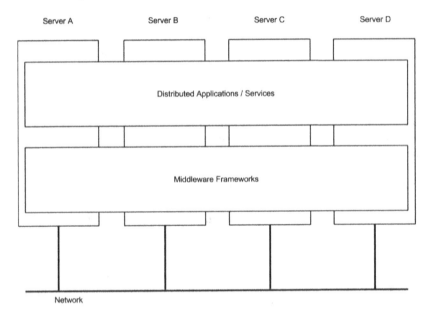

Fig. 1.1: High-level distributed system layout

corporate network systems, manufacturing network units, institution networks, home/personal networks, and in-transit networks like in buses and cars, etc., are connected. The ubiquitous nature of these networks and their growing numbers caters to many use-cases that open the barriers of distributed computing as a whole. All these networks share similar characteristics that make them perfect subjects for study under the field of distributed computing [3]. Figure 1.1 is a visual representation of the definition so far. It represents the essential characteristics where networked computers perform computations and compute with each other through a distributed service or program using software constructs that emulate distributed memory and communication interfaces.

While building Distributed Systems which may map across geographies or are connected in the same building address the following pain points:

1. **Concurrency:** Concurrent program execution is a norm while executing Distributed Systems. The challenge is to establish proper communication and minimize computer latency in executing shared resources. Concurrency becomes relevant mostly in dealing with transactional processes like the databases. Capacity of the system to handle shared resources can be increased by adding more resources (e.g., computers) to the network.

2. **No Central Clock:** When programs that are distributed across the network need to cooperate they exchange messages through a Message Passing Interface. Close coordination between the processes requires an idea of shared time to keep track

of the program state and progress. It is observed that there are limitations to the accuracy with which the nodes in the network synchronize their clocks as there is no concept of a global clock. This is the result of the fact that the only form of communication is by sending messages through a network.

3. **Independent failures:** Designers must build for failures since the most common form of disconnection is due to infrastructure failures. Computer systems often fail and the consequences of these failures must be addressed. Distributed systems can fail because of faults in the network that can isolate computers, though they do not stop running. Similarly, working nodes in the system can fail due to system crashes or unexpected termination of the processes which when not immediately made known to the other components, with which it communicates, can hamper the overall functioning. Each component of the system can fail independently, leaving the others still running. The consequences of this characteristic needs to be taken care of and are usually called fault tolerance mechanisms [3].

The goal behind creating and maintaining distributed systems is the desire to share resources. A *resource* forms a generic description of anything ranging from storage disks, printers to software-defined databases, and data objects. It also includes digital streaming be it audio or video. Over the past years, distributed systems have gained substantial importance. The reasons for their growing importance are manifold:

1. **Geographically Distributed Environment:** The most apparent idea of distributed systems is the physical distribution of computers around the globe. For example, in the case of banks that are present in multiple locations to handle the accounts of its customers. The banks are truly connected, also called globalized, if the banks can monitor interbank transactions, monitor and record fund transfer from globally dispersed ATMs, and access customer accounts around the globe, etc. Another example is the Internet that takes the nadir point in the functioning as a distributed system. Furthermore, the mobility of the users has further added a new dimension to geographic distribution.

2. **Computation Speed:.** Increased computation speed forms perhaps the most important motivation for considering distributed architectures. There is a physical limit to how much a uniprocessor can compute in unit time. While super-scala and VLIW (Very Large Scale Instruction Word) architectures provide boost to the processing power by introducing parallelism at the instruction level, these techniques do not scale well beyond a certain level. An alternative to this approach is to stack multiple processors and dividing the problem set in to smaller problems and assigning these to individual processors that can operate concurrently. This method is scalable and processing power can be gradually increased by stacking more processors. This forms a simpler and economic alternative to purchasing a single large uniprocessor. In the recent times, big data computations has led to the birth of software systems that can break the problem into smaller bits and

spread them across the network for multicore computations due to the distributed nature of the data.

3. **Resource sharing:** A *resource* represents both software and hardware. The ubiquitous nature of distributed systems is due to the resource sharing facility. A user of computer A might want to use the printer connected to computer B, or the user of computer B may want to use the extra disk storage available in computer C. Likewise, a workstation might want to use the idle computing powers of workstations B and C to speed up the current computation. These scenarios form ideal use-cases for distributed systems. Distributed databases are good examples of sharing software resources where a large file may be stored in several host machines and need to be consistently retrieved and updated by a number of processes.

   Resource sharing is a commonplace activity with networked resources today. Hardware resources such as printers, files, and more specifically to functionality such as search engines are routinely shared. At a hardware provision standpoint sharing resources like printers and disks reduce costs, but users are interested in sharing on a much higher level like their applications, everyday activities, etc. An user is interested in sharing web pages rather than the hardware it is implemented on; similarly users are more interested in sharing applications like search engines and currency converters than the servers these applications depend on. It is important to recognize that strategies to implement resource sharing vary widely in their scope depending on how the users work together; for example, a search engine serves users throughout the world, while a closed group of users collaborate directly by sharing documents.

4. **Fault-tolerance**. The software built around a single, powerful uniprocessor is prone to complete collapse if the processor fails and this is risky. It is preferable to compromise a bit on the processing by using distributed systems, when a failure cripples a fraction of the processing element with a chance for graceful degradation and improved tolerance. This method can also enhance the reliability and availability of the system by incorporating the redundant processing elements. In many cases, a system can have triple modular redundancy where three identical functional units are used to perform the same computation and the correct answer is typically determined by a majority vote. In other fault-tolerant distributed systems, processors cross-check one another at predefined checkpoints, catering to automatic failure detection, diagnosis, and eventual recovery. Thus a distributed system provides an excellent opportunity for incorporating fault-tolerance and graceful degradation.

1.2 Types of Distributed Systems

1.2.1 Distributed Embedded System

The story so far about distributed system has been largely on the physically distributed side where nodes are fixed and have a more or less permanent and high-quality connection to a network. This stability is due to the numerous techniques available that give us the impression that only occasionally things go wrong. These techniques like masking and recovering from failure, hiding the distributed aspect of the nodes effectively allows the users and the applications to believe in the reliability of these systems.

However, in recent times, several assumptions have been challenged with the growing number of mobile and embedded computing devices, which are instable at best. These devices are predominantly battery powered, mobile, and having a wireless connection. This distributed embedded system is part of our surroundings. These devices lack general administrative control, at best these devices need to be configured by the users or they need to automatically discover their environment and fit in to the network as best as possible. The following criteria will guide us with this:

- React to contextual changes

- Encourage ad hoc composition

- Recognize sharing

Reacting to contextual changes means these devices must be continuously aware of the changing environment around them. One of the simplest changes is to discover network availability.

Ad hoc composition refers to the various use cases for these pervasive systems. As a result, these devices must be configured through a suite of applications running on a device either by the user or through automated processes. These devices typically join networks to access information. This calls for a means to easily read, store, manage, and share information. The place where the information resides changes all the time; hence, the device must be able to react to the service accordingly. The distribution of data, process, and control is a property in these systems and it is better to expose these than to hide them. Here are some concrete examples of pervasive systems:

- **Home Systems:** Home networks have increasingly become the most popular type of pervasive systems. They generally consist of one or more computers in their network, but typically connect consumer electronics such as TV, audio and video equipment, gaming devices, smartphones, PDAs, and other wearables to their network. In addition kitchen appliances, surveillance cameras, clocks, controllers for lighting, and so on can also be connected to a single distributed system. Before home networks become a reality, several challenges have to be addressed. *Self-Configuring* and *Self-Managing* should be the most important feature of any home network. The networks fail if one or more devices are prone to errors. Much of these networks are addressed by Universal Plug and Play (UPnP) standards by which these devices obtain IP addresses automatically, and can discover and communicate with each other. However, more is needed to accomplish a seamless distributed system.

Some areas of concern with distributed home systems are: it is unclear how the software and firmware on the devices can be easily updated without manual intervention. Considering the unique arrangement of a home system where there are shared devices and personal devices, it is unclear how to manage the personal space where data in a home system is also subjected to personal space issues. Much of the interest in this domain is in establishing these personal spaces. Fortunately, things may become simpler. It has always been envisioned that the home systems are inherently distributed. Such a dispersion can lead to data synchronization problems. However, these problems are rapidly addressed by increasing the capacity of hard disks, along with the decrease in the size. It has become easier to configure a multi-terabyte storage unit. The size of these units have decreased such that multiple hundreds of gigabytes storage are being fit into these portable devices. This facility allows us to have a master client arrangement where there is a single system that stores and manages all the data of the network and all the clients are just interfaces to this master system. This method does not solve the problem of managing personal spaces. The Ability to store a large amount of data shifts the problem to storing relevant data and being able to find it later. In recent times, home systems come equipped with a recommender, which identify a similar taste in data stored and subsequently derive the content that is relevant to a user's personal space.

- **Health Care:** An important class of pervasive systems is based on electronic health care. With the increasing cost of medical treatment, new devices are being developed to monitor the well-being of individuals and to automatically update the concerned physicians. The goal of these systems is arranged in a Body-Area Network (BAN) which only hinders a person minimally. The network needs to be wireless and functional while the person is moving. This requirement leads to two obvious organizations, as shown in Figure 1.2. In the first arrangement, a central hub is part of the BAN and collects the data as and when generated. The data collected will be stored in a data store in a timely fashion, in this way the hub can manage the BAN. In the second arrangement, the BAN is continuously hooked to a network to which the monitored data is continuously sent. Separate techniques will be needed for managing this BAN.

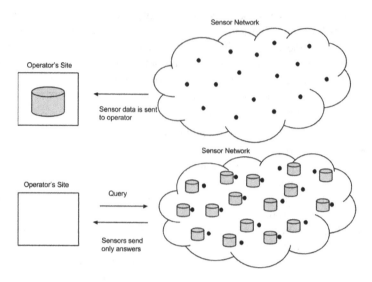

Fig. 1.2: Organising Sensor Networks [3]

- **Sensor Networks:** Our last example of pervasive systems is sensor networks. These networks in many cases form the enabling technology for many pervasive applications. What makes sensor networks interesting with respect to distributed systems' perspective is that in virtually all cases they are used for processing information.

  A *sensor network* typically consists of nodes ranging from tens to hundreds or thousands. Each node is equipped with a sensing device. Most sensor networks use wireless communication, and the nodes are often battery powered. Their limited resources, restricted communication capabilities, and constrained power consumption demand that efficiency be high on the list of design criteria.

  The relation with distributed systems can be made clear by considering sensor networks as distributed databases. In this case, an operator would like to extract information from (a part of) the network by simply issuing queries such as *"What is the northbound traffic load on Highway 1?"* Such queries resemble those of traditional databases. In this case, the answer will probably need to be provided through collaboration of many sensors located around Highway 1, while leaving other sensors untouched.

  To organize a sensor network as a distributed database, there are essentially two extremes. First, sensors do not cooperate but simply send their data to a centralized database located at the operator's site. The other extreme is to forward queries to relevant sensors and to let each compute an answer, requiring the operator to sensibly aggregate the returned answers. Neither of these solutions are very attractive. The first one requires that sensors send all their measured data through the network, which may waste network resources and energy. The second solution may also be wasteful as it discards the aggregation capabilities of sensors which would allow much less data to be returned to the operator. What is needed are facilities for in-network data processing, similar to those in pervasive healthcare systems.

In-network processing can be done in numerous ways. One obvious way is to forward a query to all sensor nodes along a tree encompassing all nodes and to subsequently aggregate the results as they are propagated back to the root, where the initiator is located. Aggregation will take place where two or more branches of the tree come together.

1.2.2 Distributed Information System

Another important class of distributed systems, found prominently in organizations, are the distributed information systems. These systems have at their core few networked applications. The server running the application is connected to the network and other systems in the network are able to talk to this server via clients. Such clients will send request to the server for execution and receive and process the server response.

Integration at a lower level allows the clients to wrap multiple requests into a single larger request and have it executed as a distributed transaction. The key idea was that all or none of the requests will be executed.

As applications became more sophisticated and were gradually separated into independent components (notably distinguishing database and processing components), it became clear that integration should take place at the component level with the components communicating with each other.

1.2.3 Distributed Computing Systems

This is a class of distributed systems that are predominant in high-performance computing tasks. One can make a general distinction between two subgroups, cluster computing systems and grid computing systems. In cluster computing, the underlying hardware consists of a collection of similar workstations or PCs, closely connected to each other via a high speed network. Each of the connected nodes runs the same operating system. This is different when it comes to grid computing systems, as the distributed systems are often constructed as a federation of computers where the system may fall into different administrative domains with different software, hardware, and network specifications.

Cluster Computing Systems

The cluster computing sphere becomes more attractive when the price-to-performance ratio of personal computers and workstations improves. It is financially and technically attractive to build a supercomputer using off-the-shelf technology by simply connecting computers over a high-speed network. In almost all cases, cluster computing is used to run parallel applications where a single program is run on all systems in the cluster [5].

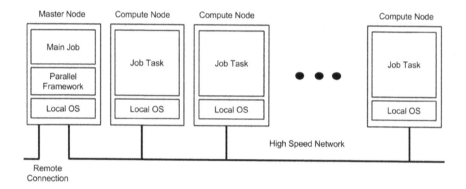

Fig. 1.3: Overview of a Cluster [5]

A well-known cluster is formed using Linux-based Beowulf cluster as shown in Figure 1.3. Each cluster consists of a collection of compute nodes/worker nodes that are accessible, controlled, and accessed by a master node. The master node typically handles the allocation of the workers on the parallel program, maintains the job queue, and provides an interface for the users to monitor these jobs. Typically, the master runs the middleware required for the execution of the program while the worker nodes just require a standard operating system.

Grid Computing

An important feature of cluster computing is homogeneity. In most cases the nodes of a cluster are similar as they have the same operating system, connected to the same network. In contrast, grid computing systems have a high degree of hetero-geneity where hardware, software, operating systems, networks, security policies, and administrative domains are largely unexplained [6]. A key challenge in grid computing is that resources from different environments are brought together to allow the collaboration of a group of people or institutions. The resources that are allocated to an organization(refers to a sample setup for collaboration) are visible to all users in that organization and this setup is usually termed virtual organi-zation. Typically, the resources consist of compute servers, storage facilities, and databases. In addition, special networked devices such as telescopes, sensors, etc., can be included as well.

The architecture consists of four layers [7] as shown in Figure 1.4. Describing the layers from the bottom, First layer, *Fabric Layer*, interfaces to the resources directly at a site. These interfaces are made to allow for resources sharing within a virtual organization. These interfaces provide functionality to query the state and capabili-ties of a resource also manages the resource functionality like resource locking, etc. The *Connectivity Layer* extends communication protocols for supporting grid trans-actions that span the multiple resource usage. In addition to the communication the layer also extends security features like authentication and authorization. In many cases programs are authenticated to access the resources rather than users. Granting these rights to specific programs is extensive when discussing security in distributed systems.

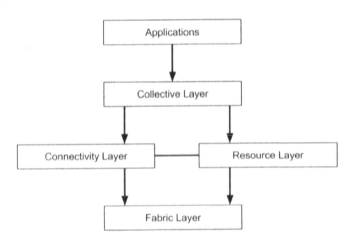

Fig. 1.4: Layered view of Grid Computing [6]

The *resource layer* manages resources. It acts as an intermediary as it uses the func-tions given but the connectivity layer calls directly the interfaces made available by the fabric layer. For instance, if you need to obtain the configuration details of a resource, creating a process to read the data from the resource. The resource layer thus extends access controls toward the resources and relies on the security and authorization rules from the connectivity layer.

The *Collective Layer* with the help of the resource layer deals with the management of multiple resources. It consists of services for resource discovery, allocation, and scheduling of tasks onto multiple resources, data replication, etc. Compared to the connectivity and resource layers it has more protocols defined for different purposes that reflect a broad spectrum of services it may offer to a virtual organization.

Lastly, the *Application Layer* represents the applications and software that are defined to take advantage of the grid computing environment.

The Collective, Connectivity, and Resource layers typically form an integral part of the grid system and are also termed as *Grid Middleware Layer* .

1.3 Distributed Computing Architecture

Distributed systems are complex software components that are, by definition, dispersed across multiple machines in the network. These systems need to be organized to be understand their complexity. There are a number of ways to visualize this organization; a one is to make a distinction between the logical organization of the collection of software components and the other is the physical distribution of the member systems. These architectures tell us how these software components are to be organized and communicated with each other [8]. There are many ways in which this can be done. The final instantiation of the software architecture can also be called *System Architecture* . Different components and connectors can be used with various configurations, which are classified into architectural styles. A few prominent ones are:

- Layered architectures
- Object-based architectures
- Data-centered architectures
- Event-based architectures

Layered Architectures: are typically uni-directional and simple, where the flow is top-down. It is organized such that the Layer L can call the underlying Layer L_i, but the reverse flow is not allowed.

The *Object-based* Architecture encourages a looser organization where the components are referred to as Objects and they are connected through Remote Procedural Calls. This type arrangement matches the client-server architectures.

Data-centered architectures include processes communicating through a common repository as a basis. This type of architecture is perhaps the most common with

Internet-based applications. A large number of applications have been built that rely on distributed file systems through which all the communication takes place. Likewise, most web-enabled distributed systems are data-centric where processes communicate with each other through the use of shared web-based data services.

In *Event-based* the processes are designed to communicate through event handling. Occasionally these events carry data. Typically, distributed systems like publish/-subscribe are built using even-based principles. The typical idea behind this style is that the processes publish events at different stages of execution and the middleware ensures that the processes that are subscribed to these events receive them. One of the many advantages is that the processes are loosely coupled as they do not depend on each other explicitly. This is also described as being referentially decoupled.

Event-based architectures can be combined with data-centered architectures, yielding what is also known as shared data spaces. The essence of shared data spaces is that the processes are now also decoupled in time: they need not both be active when communication takes place.

1.4 Distributed File Systems

Sharing of resources is a key goal for distributed systems. Sharing stored information is an important aspect of distributed resource sharing. There are many mechanisms for data sharing. The files are stored locally in web servers, in file systems at the server, or in servers on a local network are made available to clients throughout the Internet. Designing large-scale read-write file storage systems over the Internet has several problems like load balancing, reliability, availability, and security. Replicated storage systems are suitable for applications requiring reliable access to data stored on systems where the availability of individual hosts cannot be guaranteed.

Sharing data and information within local networks requires persistent storage of data and programs of all types with consistent data distribution. Originally, file systems were developed for centralized computer systems and desktops where an operating system provides convenient access to disk storage. Soon they acquired features such as access-control and file-locking mechanisms that are necessary to execute sharing of data and programs. A well-designed distributed file system provides file access stored on multiple servers with performance and reliability similar to files stored on local disks. Their design is adapted to the performance and reliability characteristics of local networks.

1.4.1 DFS Requirements

Distributed File Systems development has been a starting point in identifying many of the challenges and pitfalls in designing distributed services. The development realized access transparency and location transparency in the earlier phases, performance, scalability, concurrency control, fault tolerance, and security requirements were met in the subsequent phases of development.

1. **Transparency:** The file systems receive maximum load in any distributed arrangement, hence its functionality and performance are very critical. The design of the file service must balance scalability and flexibility due to transparent software complexity and performance. The following forms of transparency are:

   - *Access Transparency:* The distribution of files should be abstracted from the client programs. APIs are provided with programs to modify local and remote files easily.

   - *Location Transparency:* Since the data are replicated and distributed on different computers it is important to have a uniform name space. The files may be relocated without the need to change their pathname. It is important that the client programs see the same name space anywhere in the network.

   - *Mobility Transparency:* The client programs and system tables need not be changed when the files are moved. This allows file mobility.

   - *Performance Transparency:* The client programs should continue to provide the expected performance when the load on the system varies within a range.

   - *Scaling Transparency:* The distributed services can be expanded or reduced based on the load and network sizes.

2. **Concurrent File Updates:** File updates by one client should not interfere with other clients simultaneously accessing or changing the same file. This is known as concurrency control. Concurrency control is a widely accepted facility and there are several techniques known to implement this. Most UNIX-based systems provide block level or file level locking.

3. **File Replication:** A file or block of data is represented by several copies in different locations. This has two benefits - First, it enables multiple servers to share the load of providing a service accessing the same file. Secondly, this method scales to multiple requests and enhances fault tolerance by locating another server that holds a copy of the requested file when one server fails. Few file services have full replication while many use caching of files or replicating portions of files.

4. **Hardware Heterogeneity:** The services interfaces should be defined such that they can be implemented on different operating systems. This forms the basis for openness.

5. **Fault Tolerance:** The ability of the service to continue working in the event a client of server fails. Several fault-tolerant mechanism are employed like at-most-once invocation or at-least-once with a server ensuring that duplicated results do not result in invalid updates to the files. The servers can be stateless, so that they can be restarted and restored after a failure to recover the previous state.

6. **Consistency:** Conventional file systems provide UNIX style one-copy update schemes where the file updates are handled as if there were only one copy of the file and the updates are propagated to all the clients viewing the file. In the case of distributed file systems where there exists multiple copies of the file, synchronizing the changes will take time which may result in data consistency issues.

7. **Security:** In distributed systems there is a need to authenticate the user to protect the contents of the request and sometimes reply with digital signatures and encrypted data. Access control lists are used to execute this.

8. **Efficiency:** The distributed systems must provide services that are of the same or, in general, comparable level of performance as that of conventional file systems.

Techniques used for implementation of file services are important in distributed system design. In general the distributed system must provide a service comparable to the conventional file services. It must be convenient to administer, providing operations and tools that enable system administrators to install and operate the system conveniently.

1.4.2 DFS Architecture

Most systems are built following a traditional client-server architecture, but fully decentralized solutions exist as well. This section explains both kinds of organizations.

- **Client-Server Architectures:** Client-server architecture is the most common way of organizing distributed architectures. Sun Microsystems Network File System(NFS) is one of the most widely-deployed file services for UNIX-based systems. The basic idea behind NFS is that each file server provides a standard view of the file system. In other words, it does not matter how the file system is implemented, each file server supports the same model. This model is popular in other implementations as well. The NFS provides its own communication protocol that allows clients to access the files, thus enabling heterogeneous processes running on different operating systems and hardware to share a common file system.

  The clients are unaware of the location of the files stored in a distribute file service. Instead, an interface is provided to the file system that is similar to the conventional local file system. For example, Hadoop provides command line interface to copy, delete, and list files in the Hadoop Distributed File System(HDFS). These interfaces can be used for various operations that are implemented by the file server specifically. This model is similar to a remote file service and hence is called the remote access model [11].

In another model, Upload/Download model, a client can access a file by downloading it, after which he/she makes the changes and it is uploaded back to the file server so that it can be used by other clients. An important advantage with this model is that it does not matter what the file system on the client side is, whether UNIX, Windows, or even MS-DOS, all that matters is the file systems are compliant with the file system model offered by the NFS.

- **Cluster-Based Distributed File Systems:** Though NFS is a popular architecture for client server-based distributed system, it is enhanced for server clusters with a few differences. Considering the parallel applications of the clusters the file systems are also adjusted accordingly. A common technique is to implement file-striping, by which a single file is distributed across multiple servers. The idea behind this is fairly straightforward, if the file is distributed across several nodes then the parts of the files can be retrieved in parallel. This can work only if the application is organized such that parallel data access makes sense. This requires that the data stored have a regular structure, for example, dense matrix. For other structures file-striping may not be effective. In these cases it is better to partition the file systems themselves instead of the files, and store the files individually in different partitions.

This task of distribution becomes interesting when distributing data across large data centers in companies like Amazon and Google. These companies offer services resulting in reads and updates of a massive number of files distributed across tens of thousands of computers [10]. In these cases the traditional distributed file system assumption does not hold because at any moment there will be a computer that can fail. To counter these problems, Google, for example, has developed its own file system called Google File System(GFS) [12].

The GFS cluster consists of a single master along with multiple workers called chunk servers. Each file is divided into chunks of 64 Mbyte each. These chunks are distributed across chunk servers. It is important to note that the GFS master is contacted only for meta-data information, expecting a contact address for a chunk. The contact address has all the information about the chunk server and the chunk it carries. The master maintains the name space along with the mapping of the file name to chunks. Each chunk has an identifier associated with it that will allow the chunk server to look it up. The master also keeps track of where the chunk is located. These chunks are replicated to handle failures, but no more than that. The master does not keep an accurate chunk location, instead it occasionally contacts the chunk servers and requests for chunk locations. This is done periodically, otherwise the master would become complicated if you need the master record to be consistent because every time a chunk server crashes or a new server is added the master needs to be informed. Instead of this it is easier to update the current set of chunk servers through polling. The GFS clients get to know which chunk servers the master believes is storing the requested data. Since chunks are replicated, there is a high probability that a chunk is available on at least one of the chunk servers.

Is this solution scalable? An important design choice is that the master is largely in control, but that it does not form a bottleneck due to all the work it needs to do. Two ways to accommodate for scalability are :

- Majority of the work is done by the chunk servers. When a client needs to access data, it requests the master for the address of the chunk server with the required data after which it contacts the chunk server directly. Chunks are replicated according to a backup scheme. When the client performs an update operation it contacts the nearest chunk server holding the data and pushes the changes directly to it. Once the update is successful a sequence number is allocated to the update and it is passed on to all the backups. The master is kept out of loop, removing the master bottleneck for every update.

- The hierarchical name space for the files is implemented using a single level table in which the file paths are mapped to the meta-data (equivalent to i-nodes in traditional file systems). This single table is kept in the main memory along with the mapping of file chunks. Updates to these chunks are logged to persistent storage. When the log becomes too large a checkpoint is created on the data in the main-memory so that it can be immediately restored back in the main-memory. Consequently, the I/O of the GFS cluster is reduced significantly. This organization allows a single master to control a few hundred chunk servers, which is a considerable size for a single cluster. By subsequently organizing a service such as Google into smaller services that are mapped onto clusters, it is not hard to imagine that a huge collection of clusters can be made to work together.

1.5 Challenges in Distributed Systems

Several challenges extend the scope and scale of distributed systems. Some of the challenges are:

Heterogeneity: The Internet is a heterogeneous arrangement of a distribute system that is used to run services and applications. Heterogeneity applies to the following:

- Networks
- Computer hardware
- Operating systems

- Programming languages

- Implementations by different developers.

It can be observed that the Internet has several types of networks connected to each other, they communicate with each other using the Internet protocols implemented on each of the connected computer. This common interface masks the differences in these networks. If a computer is connected to the Internet over the Ethernet the Internet Protocols need to be implemented over the Ethernet.

Data types such as integers can be represented differently on different hardware - example, there are two ways for byte ordering of integers. These differences must be addressed if there is to be communication between programs running on different hardware. Although the operating systems on all the computers with different hardware have the implementation of the Internet protocols they do not provide the same Application Programming Interface, the APIs are different for Unix and Windows.

To further discuss inter program communication it is vital to define a standard for programming network applications. The programs written by developers cannot communicate with each other unless they use a common standard. Like the Internet protocols which are nothing but agreed upon standards, there is a need for standards in programming applications that survive on networks.

Openness: A system's *openness* is the ability to extend or reimplement that system. The openness with respect to distributed system is determined by the ability to add more resources that can be shared to various client programs.

Openness can be achieved as long as the specification and documentation of key software interfaces are made available to developers. This process is similar to standardization procedures but usually less cumbersome compared to the general method. However, documentation and specification publications of interfaces are only the beginning to extend services; the real challenge is the complexity of the distributed systems that consist of many components built by different people.

The systems that are designed for resource sharing with open documentation are extensible. These can be extended at the hardware level by adding more computers and at the software level by the introduction of new services or reimplementing the old ones. Another added benefit is they are independent of vendors due to the openness [9].

Some of the characteristics are:

- An important characteristic is the open documentation of the interfaces.

- Open distributed systems provide a uniform communication mechanism and published interfaces for access to shared resources.

- Open distributed systems can be built from heterogeneous hardware specifications and software from different vendors but, it is vital to conform that each component is working according to the published standard.

Security: Information security has great importance with distributed systems as the information is also distributed across multiple information storage resources in the system. The security identifies three components:

- *Confidentiality:*, Data protection and prevention of unauthorized disclosure.

- *Integrity:* Protection against the corruption of data.

- *Availability:* Protection against any interference to access the stored information.

There is always a security with free access given to information on the Internet. Although a firewall could relatively form a barrier restricting traffic around the Internet this does not, however, deal with the appropriate use of the resources by users within the Internet that are not protected by the firewall.

In a distributed system a client sends requests to access the data hosted by the servers, these data are packed as responses back to the client over a network. Some examples are:

- Health Care: A doctor might request or update a patient data with respect to the progress of the patient's ailment.

- E-commerce: Ability to purchase products online requires you to send confidential credit card information over the Internet.

In these above scenarios the challenge is to send the sensitive information across a network with security as paramount. The security not only includes protection of the message but also authentication of the user/agent on whose behalf a message was sent. In the first case the credentials of the doctor need to be verified before any transmission of messages. In the second case the authorization of the remote agent needs to be verified. One way to approach these problems is addressed by using encryption techniques developed for this purpose [13].

Denial of service attacks: This is one of the most common security threats in recent times which can be used to disrupt any service over the Internet. This can be achieved by flooding the server with a lot of requests such that this service is busy for the real user. Hence the name denial of service attack. Currently such attacks are punished after such an event but does not provide an opportunity to counter this problem before the attack [14].

Scalability: Scalability is rather a desirable outcome as distributed systems operate at different scales from small Intranets to Internet. Scalability is achieved when a distributed system can perform as per requirements with increase or decrease in the number of resources and the number of users.

It is interesting to note that the number of web servers increase in this period and also the quick saturation which can be explained by the trend to increase the capacity of the distributed system serving the dramatic growth of mobile devices. Also, there is a web server being hosted in multiple locations to increase availability. The design of scalable distributed systems presents the following challenges:

1. **Costing of physical resources:** Main advantage of using distributed system is the ability to extend the system as when the demand for resources increases. Resource addition must be done at a reasonable cost. The need for resources increases when there is an increase in the number of users. Distributed systems allows of adding new computers to avoid performance bottlenecks as in case of a single server setup. A general, but not limited to, principle is if there are n users and for the system to scale the quantity of resources can be utmost $O(n)$, proportional to n. In most cases it is always several magnitudes lesser.

2. **Limiting performance loss:** In a system that contains data proportional to the number of users, for example, Internet address looks up database, i.e., the addresses of the computers and their domain names. Algorithms that use hierarchic structures to store this data is more scalable compared to the linear system. The access time with hierarchic structures is $O(\log n)$, n is the size of data. For a system to be scalable the performance should not be worse than $O(\log n)$.

3. **Prevent resource scarcity:** A scenario where scalability is an issue is the number of available Internet addresses for the systems. Earlier 32-bit addressing was used and which recently has reached it maximum limit and now 128-bit addressing protocol is being defined and used, but this adoption requires change in many Internet facing software components. To be fair, the designers do not have a solution for this problem but to handle the requirement as it arrives, so far. It is easier to adapt to the change in requirements than to overcompensate for it as larger Internet addresses will use more space in messages that are sent to and fro in a network.

4. **Preventing performance bottlenecks:** Algorithms need to be decentralized to avoid performance bottlenecks. A very good example is with the Domain Name System(DNS). Early DNS management included a single master file with the Domain Table. This file was downloaded on request, this method is sufficient for a few hundred system which soon started to become a performance issue with new computer being added. The DNS management system removed a single access point and partitioned this data across the Internet and maintained it locally.

Failure Handling: Computers fail, whether it is hardware malfunction or a software fault. Programs fail to complete their intended purpose because of failures. There are number of failure possibilities that can occur in the processes and networks that are a part of the distributed systems. It is unique with distributed systems because the failure of the components in this system only renders partial failure which leads to difficulty in finding the failure and fixing it.

Detecting failures: Some failures can be detected. For example, you can use checksums to detect any corrupted data in a message or a file. Some failures cannot be detected as a result these can be suspected and provisions can be made to manage them. Two examples of hiding failures:

- Message retransmission when it fails to arrive.

- Duplication and Replication data to account for any corruption of disks.

Although these methods are just attempts to handle failure it does not guarantee to work in the worst cases, where replicated disk also fails and network downtime could lead to the message not being transmitted.

Tolerating failures: Most of the services in the Internet are prone to failures and it is impractical to handle and manage these failures that range across many components. Sometimes it is better to handle the failures on the client side then it would mean the users tolerate these failures too. For example, when a web browser cannot connect to a web server it does not make the user wait until the connection is available rather it informs the user about the problem leaving them free to try again later.

Recovery from failures: Recovery from failure requires the software design to include data *rollback* facility. This feature enables the data to be backed up and in case a server crashes the data on the server can be restored to the previously saved state. In the event a failure occurs the computations would not have completed and the changes these would have made on the data would not be consistent, hence a state with the most consistent data is recovered.

Redundancy: Redundancy is an important strategy to manage failures. Consider the following examples:

- There should always be at least two different routes between any two routers in the Internet.

- In the Domain Name System, every name table is replicated in at least two different servers.

- Replication of data on multiple servers ensures that data remain accessible even if one of the servers fail. Servers can be designed to detect faults in their peers; when a fault is detected, clients are redirected to the remaining server.

Quality of Service (QOS): Service quality can be requested after the delivery of user required services. Reliability, security, and performance are a few non-functional properties that affect the quality of service of any product. Furthermore, adaptability to the changing system configuration and availability of resources is an important aspect of service quality.

Service security and reliability are critical to the design of distributed systems. Initially performance of a service was defined in terms of responsiveness and computational throughput, but in the recent times replication of data on multiple servers ensures data remain accessible even if one of the servers fail. Servers can be designed to detect faults in their peers; when a fault is detected clients are redirected to the remaining servers.

Mulimedia applications handle time critical data such as streams of data that need to be processed and transfered between programs at a fixed rate. A scenario where a client program receives video stream from the video server. For better user experience, successive frames of video need to be displayed to the user within specified time limits.

The ability of the systems to meet such deadlines depend on the availability of the necessary computing and network resources at appropriate times. This is rightly termed as Quality of Service(QoS). In-order for these systems to provide a better service quality, the right computational and network resources should be made available on time.

With respect to networks, today they have grown to be more reliable to provide high performance but when these are heavily loaded the performance takes a hit. Similarly with respect to individual systems in the Distributed arrangement the resources need to be managed for the desired QoS, operating systems come equipped with resource managers that keep the QoS in check. Resource managers have the ability to reserve the resources and reject a request when they are not available.

1.6 Trends in Distributed Systems

Over a period of time distributed systems have undergone changes. There are several reasons for changes in technology and the burgeoning growth of distributed tools. Few are mentioned below:

- Emerging pervasive networking technology
- Emerging ubiquitous computing coupled with the desire to support user mobility in distributed systems
- Rise of multimedia services
- Utility based distributed system

Major trends and changes can be observed in the following areas:

Emerging Network Technologies: The Internet today is a collection of vast interconnected computer systems of different types. The types range from wired connections to the latest wireless communication technologies like WiFi, 3g, 4g. As a result there are wide variety of systems and devices that can be connected to the Internet at any time in any place. The programs on these devices implement the Internet protocol and establish a common communication interface for exchanging messages. The Internet is a largest distributed system. It enables the users to access services like World Wide Web, File Transfer, email, etc. The set of services available is open-ended as new services can be added by adding new computer systems to the Internet. The Internet consists of many networks and subnetworks(Intranets) that are operated by multiple corporates and organizations. These networks are protected by firewalls. A Firewall protects an Intranet by preventing unauthorized messages from leaving and entering. A firewall filters incoming and outgoing messages. The filtering can be done either at the source or the destination, or the firewall might allow only email, web access to pass through.

The Intranets are linked together by backbones. A backbone is a network link with a high transmission capacity, employing satellite connections, fiber optic cables, and other high-bandwidth circuits. The implementation of the Internet and the services that it supports has led to the development of many solutions to issues in distributed systems.

Ubiquitous Computing: Rapid advancement in mobile device technologies and wireless technologies has led to the increasing integration of small and portable computing devices. These devices include:

- Laptop computers.
- Hand-held devices, including mobile phones, smart phones, GPS-enabled devices, pagers, personal digital assistants (PDAs), video cameras and digital cameras.
- Wearable devices, such as smart watches with functionality similar to a PDA.
- Devices embedded in appliances such as washing machines, hi-fi systems, cars, and refrigerators.

The portability and their ability to connect to wireless networks easily makes mobile computing possible. Mobile computing is computing on the move where the devices can access the information from either their home Intranets or from the Internet easily through wireless communication. Resources such as printers can be used remotely using mobile devices. Mobile computing opens new fields of research that includes location-aware and context-aware computing.

The term ubiquitous in Ubiquitous Computing tends to convey that small mobile devices will become increasingly popular so that they will be scarcely noticed. This allows for harnessing small, cheap computational capabilities of the users devices that are in currents users' environment. The computational behavior will be a function of their physical location.

Ubiquitous computing truly gets its value only through communication of the devices in the network. For example, it is convenient for users to control their washing machines or their entertainment systems from their phones. Similarly, washing machines can signal back to the user about the status of the wash.

The user can use three forms of wireless communication for his needs. First, using wireless LAN that can extend up to 100m (a building floor). This LAN is connected to the Internet via an access point or gateway. Secondly, using mobile(cellular) tele-phone, which is connected to cellular towers, in-turn connected to the Internet. The mobile phones can use this to access information on the Internet and sometimes even communicate the location of the device using the built in GPS module. Finally, using ad-hoc networks either at home or work, the mobile devices can be connected to a personal wireless network with a device such as a printer.

Supported by a suitable architecture, the users can use mobile devices to browse through information on web sites on the move, calculate the route to a destination using the built-in GPS, etc. The users can print out documents and photos directly from the phones and also project these using a projector connected to the local wireless network.

Utility Computing: In recent times, distributed computing resources are viewed more as utilities which the users can rent and use on demand instead of owing the infrastructure similar to other utilities like water and electricity. This has led to newer areas of business. The utility model applies to both Physical resources and Logical services in the network:

- **Physical resources** like storage and computing can be made available to users on demand without the need for them to own. Sometimes a user can request for a facility to store or backup their multimedia data like music, photographs or video. The service can be tailored to provide storage with some computational capability. In other cases, an user can request for data center to store and process big data with volumes and computational needs that companies like Amazon and Google can provide for distributed processing needs.

  Virtualization forms the basis for this extension of resources on demand, imply-ing the resources can be made available virtually rather than a physical node. This offers greater flexibility to the service provider in terms of resource management.

- **Software services** can also be made available across the Internet on demand. Many companies offer services like email and calendars on rental. Google, for example, bundles a range of business services under the banner *Google Apps*. This development is enabled by agreed standards for software services

The term Cloud Computing is used to represent distributed computing as an util-ity. A cloud forms a blanket over a set a Internet based application, storage, and computing services enough to support most users' needs. This allows the users to dispense their local storage and computing resources. The cloud also promotes the uses of any resource as a service be it software or hardware. The effectiveness of the cloud stems from the easy to sell pay-per-use model. The Cloud also has reduced the resource requirements from the users side and encourages the users to use small mobile devices to leverage greater computational and storage services on demand (Figure 1.5).

Fig. 1.5: Cloud computing as an utility [8]

Clouds are implemented on a cluster of computers that are interconnected and cooperate closely to provide a single, high-performance computing capability of scale. It consists of commodity hardwares running standard operating systems such as Linux interconnected by a local network. Companies like HP, Sun and IBM offer high end blade solutions. Blade servers usually contain minimal computational and storage capabilities. A blade system consists of many blade servers. Due to this many blade servers can be produced at much lower costs compared to commodity computers. The overall goal of cluster computers is to provide a range of cloud services, including high-performance computing capabilities, mass storage (for example through data centres), and richer application services such as web search (Google, for example relies on a massive cluster computer architecture to implement its search engine and other services).

1.7 Examples of HPDC Systems

The ubiquitous nature of networks and the distributed services they offer are an integral part of our daily lives like Internet, World Wide Web, web search, online gaming, email, social networks, eCommerce, etc. The field of distributed systems has seen some of the most significant technological developments of recent years.

Gaining an understanding of the underlying technology nuances is central to learning modern computing.

Distributed computing systems are increasingly used in several markets solving large data problems. Some of the areas that have a good presence of distributed systems are:

Finance and commerce: e-commerce market led by giants Amazon and Ebay has made purchasing easier and increased the purchasing capabilities of users. The payment technologies have been overhauled by services like PayPal, GoogleWallet, etc. The Financial trading markets and online banking solutions have also led to the Internet financial revolution [15].

Information society: Ever increasing global information repository the World Wide Web; rise in web search services like Google search and Microsoft Bing; Emerging digitization of information from books(for example, Google Books). Rapidly increasing user generated content on media and information repositories like Youtube, Wikipedia, and Flickr; Social networking services like Facebook, Myspace, Twitter harboring user related content [16].

Creative industries: Gaming industry has been bolstered by the rise in online games which are more interactive and user tailored. Streaming Music and Movie content either through networked media devices at home or the Internet. User generated content on services like Youtube has seen as an expression of user creativity. This creativity has been encouraged by networked technologies.

Healthcare: The growth of health related informatics including storing patient data and related privacy issues; Telemedicine to deliver diagnosis and treatment remotely, also supporting advanced telemedicine services like remote surgery; Increasing networking and embedded systems solutions for assisted living, for example, monitoring the health of elderly at their homes and keeping the doctor informed of events automatically[17]; Supporting collaboration between health teams.

Education: Emerging e-learning tolls like virtual classrooms, Massive Open Online Courses (MOOCs); Support for distance learning; support for collaborative learning across different geographies.

Transport and logistics: Location based services using GPS to find routes and more generic traffic management systems; Driverless cars for a complex example of a distributed system; The rise in web and mobile based mapping services like Google maps, Nokia maps, OpenStreetMap and Google Earth.

Science: Applying the principles of distributed systems like Grid and Cloud computing to solve complex scientific problems. These services offer solutions to storing and processing large scientific data. Distributed computing has also helped increase the collaboration of scientific teams worldwide.

Environmental management: Combining sensor technology and distributed systems eases monitoring and managing of the natural environment. These are useful in areas of disaster detection applications such as earthquakes, floods or tsunamis and to coordinate an emergency response; collating and analyzing of global environmental parameters to better understand complex natural phenomena such as climate change.

Web: With over 10 billion search requests per month, Web search has emerged as a major growth industry in the last decade. The Web Search Engine should index the entire contents of the Web containing various information schemes including web pages, multimedia, books, etc. This is a complex task as the information generated keeps growing, current stats of the web see 63 billion web pages, 1 trillion web addresses, each growing by the minute. The search engine performs sophisticated processing on this huge dataset of web content. This task stands as a major challenge for Distributed Computing researchers. Google, the market leader in web search technology, has put significant effort into the design of a sophisticated distributed system infrastructure to support search. They have implemented one of the most complex distributed systems installations in the history of computing:

- A lager number of connected computers form the underlying physical infrastructure located at data centres all around the world;

- A proprietory distributed file system designed to support very large files. This file system has been optimized to have reading files at high and sustained rates);

- Supported by a structured distributed storage system that offers fast access to very large datasets;

- A locking service offers distributed locking and agreement facilities;

- A programming model that supports the management of very large parallel and distributed computations across the underlying physical infrastructure.

Network file server: A local area network consists of many computers connected via high speed links. In many LANs, usually, a separate machine is allocated to act as a file server. when a user requests access to a file the operating systems on that file server authenticates this request and checks for authorization and subsequent success or fail is handled. The user process that requests the file and the server process must work together to execute file access/transfer protocols.

Banking network: Distributed ATM networks enable an account holder to withdraw money from different accounts setup in different cities. Due to the latest agreement with the banks you can use an ATM card of one bank accout in an ATM of another bank. These debits are registered by the respective banks and the balance is recomputed.

Peer-to-peer networks: The development of P2P networks can be attributed to the success of Napster. The Napster system used to distribute music mp3 from users' computers directly instead of hosting on their own. The users form and network between them and share bandwidth to download files. The Napster used a central repository to store the file directory listing. Subsequent systems like Gnutella avoided the use of central server or central directory. P2P networks have since been applied in areas beyond sharing files. For example, the UCB created an oceanstore project, a data archiving mechanism on top of the underlying P2P network.

Process control systems: Networks are also used extensively in industrial plants to oversee production and maintenance processes. Consider an arrangement where a controller maintains the pressure of a chamber at 200 psi, As the vapor pressure increases, the temperature has a tendency to increase so there is another controller that controls the flow of the coolant. This coolant ensures that the temperature of the chamber does not exceed 200 deg F. This is a simple example of a distributed system that maintains an invariance relationship on system parameters monitored and controlled by independent controllers.

References

1. Stevens, Richard. *UNIX Network Programming, Volume 2, Second Edition: Interprocess Communications*. Prentice Hall, 1999. ISBN 0-13-081081-9
2. Patterson, David A. and John L. Hennessy (2007). *Computer architecture : a quantitative approach, Fourth Edition*, Morgan Kaufmann Publishers, p. 201. ISBN 0-12-370490-1.
3. Jean Dollimore, George Coulouris, and Tim Kindberg, *Distributed Systems: Concepts and Design*, Addison-Wesley Publishers Limited 1988
4. Sukumar Ghosh, *Distributed Systems: An Algorithmic Approach*, Taylor & Francis Group, LLC, 2007.
5. Buyya, Rajkumar. *High performance cluster computing*. New Jersey: F'rentice 1999.
6. Berman, Fran, Geoffrey Fox, and Anthony JG Hey, eds. *Grid computing: making the global infrastructure a reality*. Vol. 2. John Wiley and sons, 2003.
7. Joseph, Joshy, Mark Ernest, and Craig Fellenstein. *Evolution of grid computing architecture and grid adoption models*. IBM Systems Journal 43.4 pp. 624-645, 2004.
8. Lampson, Butler W., Manfred Paul, and Hans Jrgen Siegert. *Distributed systems-architecture and implementation, an advanced course*. Springer-Verlag, 1981.
9. Tanenbaum, Andrew, and Maarten Van Steen. *Distributed systems*. Pearson Prentice Hall, 2007.
10. Barroso, Luiz Andr, Jeffrey Dean, and Urs Holzle. *"Web search for a planet: The Google cluster architecture."* Micro, Ieee 23.2 pp. 22-28, 2003.
11. Bernstein, Philip A. *"Middleware: a model for distributed system services."* Communications of the ACM 39.2 pp. 86-98, 1996.
12. Ghemawat, Sanjay, Howard Gobioff, and Shun-Tak Leung. *"The Google file system."* ACM SIGOPS Operating Systems Review. Vol. 37. No. 5. ACM, 2003.

13. Thain, Douglas, Todd Tannenbaum, and Miron Livny. *Distributed computing in practice: The Condor experience.* Concurrency and Computation: Practice and Experience 17.24 pp. 323-356, 2005.
14. Mirkovic, Jelena, et al. *Internet Denial of Service: Attack and Defense Mechanisms* (Radia Perlman Computer Networking and Security). Prentice Hall PTR, 2004.
15. Sarwar, Badrul, et al. *"Analysis of recommendation algorithms for e-commerce."* Proceedings of the 2nd ACM conference on Electronic commerce. ACM, 2000.
16. Dikaiakos, Marios D., et al. *"Cloud computing: distributed internet computing for IT and scientific research."* Internet Computing, IEEE 13.5 (2009): 10-13.
17. Coiera, Enrico. *"Medical Informatics, the Internet, and Telemedicine."* London: Arnold (1997).

Chapter 2
Getting Started with Hadoop

Apache Hadoop is a software framework that allows distributed processing of large datasets across clusters of computers using simple programming constructs/models. It is designed to scale-up from a single server to thousands of nodes. It is designed to detect failures at the application level rather than rely on hardware for high-availability thereby delivering a highly available service on top of cluster of commodity hardware nodes each of which is prone to failures [2]. While Hadoop can be run on a single machine the true power of Hadoop is realized in its ability to scale-up to thousands of computers, each with several processor cores. It also distributes large amounts of work across the clusters efficiently [1].

The lower end of Hadoop-scale is probably in hundreds of gigabytes, as it was designed to handle web-scale of the order of terabytes to petabytes. At this scale the dataset will not even fit a single computer's hard drive, much less in memory. Hadoop's distributed file system breaks the data into chunks and distributes them across several computers to hold. The processes are computed in parallel on all these chunks, thus obtaining the results with as much efficiency as possible.

The Internet age has passed and we are into the data age now. The amount of data stored electronically cannot be measured easily; IDC estimates put the total size of the digital universe at 0.18 Zetabytes in 2006 and it is expected to grow tenfold by 2011 to 1.8 Zeta-bytes [9]. A Zetabyte is 10^{21} bytes, or equivalently 1000 Exabytes, 1,000,000 Petabytes or 1bn Terabytes. This is roughly equivalent to one disk drive for every person in the world [10]. This flood of data comes from many sources. Consider the following:

- The New York Stock Exchange generates about one terabyte of trade data per day.

- Facebook hosts approximately 10 billion photos, taking up one petabyte of storage. Ancestry.com, the genealogy site, stores around 2.5 petabytes of data.

- The Internet Archive stores around 2 petabytes of data, and is growing at a rate of 20 terabytes per month.

© Springer International Publishing Switzerland 2015
K.G. Srinivasa and A.K. Muppalla, *Guide to High Performance Distributed Computing*, Computer Communications and Networks, DOI 10.1007/978-3-319-13497-0_2

- The Large Hadron Collider near Geneva, Switzerland, produces around 15 petabytes of data per year.

2.1 A Brief History of Hadoop

Hadoop was created by Doug Cutting, the creator of Apache Lucene, a widely used text search library. The Apache Nutch project, an open source web search engine, had a significant contribution to building Hadoop [1].

Hadoop is not an acronym; it is a made-up name. The project creator, Doug Cutting, explains how the name came about:

> The name my kid gave a stuffed yellow elephant. Short, relatively easy to spell and pronounce, meaningless, and not used elsewhere: those are my naming criteria. Kids are good at generating such.

It is ambitious to build a web search engine from scratch as it is not only challenging to build a software required to crawl and index websites, but also a challenge to run without a dedicated operations team as there are so many moving parts. It is estimated that a 1 billion page index would cost around $500,000 to build and monthly $30,000 for maintenance [4]. Nevertheless, this goal is worth pursuing as Search engine algorithms are opened to the world for review and improvement.

The Nutch project was started in 2002, with the crawler and search system being quickly developed. However, they soon realized that their system would not scale to a billion pages. Timely publication from Google in 2003, the architecture of Google FileSystem, called the GFS came in very handy [5]. The GFS or something like that was enough to solve their storage needs for the very large files generated as part of the web crawl and indexing process. The GFS particularly frees up the time spent on maintaining the storage nodes. This effort gave way to the Nutch Distributed File System (NDFS) .

Google produced another paper in 2004 that would introduce **MapReduce** to the world. By early 2005 the Nutch developers had a working MapReduce implementation in Nutch and by the middle of that year most of the Nutch Algorithms were ported to MapReduce and NDFS.

NDFS and MapReduce implementation in Nutch found applications in areas beyond the scope of Nutch; in Feb 2006 they were moved out of Nutch to form their own independent subproject called Hadoop. Around the same time Doug Cutting joined Yahoo! which gave him access to a dedicated team and resources to turn Hadoop into a system that ran at web-scale. This ability of Hadoop was announced that its production search index was generated by the 10,000 node Hadoop cluster [6].

In January 2008, Hadoop was promoted to a top level project at Apache, confirming its success and its diverse active community. By this time, Hadoop was being used by many other companies besides Yahoo!, such as Last.fm, Facebook, and the New York Times.

The capability of Hadoop was demonstrated and publicly put at the epitome of the distributed computing sphere when The New York Times used Amazon's EC2 compute cloud to crunch through four terabytes of scanned archives from the paper converting them into PDFs for the Web [7]. The project came at the right time with great publicity toward Hadoop and the cloud. It would have been impossible to try this project if not for the popular pay-by-the-hour cloud model from Amazon. The NYT used a large number of machines, about a 100 and Hadoop's easy-to-use parallel programming model to process the task in 24 hours.

Hadoop's successes did not stop here, it went on to break the world record to become the fastest system to sort a terabyte of data in April 2008. It took 209 seconds to sort a terabyte of data on a 910 node cluster, beating the previous year's winner of 297 seconds. It did not end here, later that year Google reported that its MapReduce implementation sorted one terabyte in 68 seconds [8]. Later, Yahoo! reported to have broken Google's record by sorting one terabyte in 62 seconds.

2.2 Hadoop Ecosystem

Hadoop is a generic processing framework designed to execute queries and other batch read operations on massive datasets that can scale from tens of terabytes to petabytes in size. HDFS and MapReduce together provide a reliable, fault-tolerant software framework for processing vast amounts of data in parallel on large clusters of commodity hardware (potentially scaling to thousands of nodes).

Hadoop meets the needs of many organizations for flexible data analysis capabilities with an unmatched price-performance curve. The flexibility in data analysis feature applies to data in a variety of formats, from unstructured data, such as raw text, to semi-structured data, such as logs, to structured data with a fixed schema.

In environments where massive server farms are used to collect data from a variety of sources, Hadoop is able to process parallel queries as background batch jobs on the same server farm. Thus, the requirement for an additional hardware to process data from a traditional database system is eliminated (assuming such a system can scale to the required size). The effort and time required to load data into another system is also reduced since it can be processed directly within Hadoop. This overhead becomes impractical in very large datasets [14].

The Hadoop ecosystem includes other tools to address particular needs:

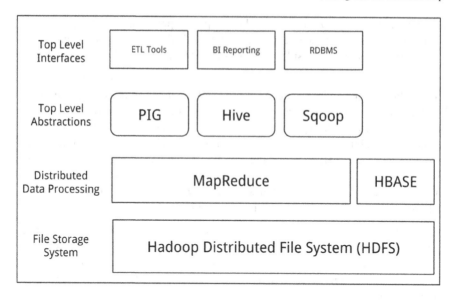

Fig. 2.1: Hadoop Ecosystem [14]

Common: A set of components and interfaces for distributed filesystems and general I/O (serialization, Java RPC, persistent data structures).

Avro: A serialization system for efficient, cross-language RPC and persistent data storage.

MapReduce: A distributed data processing model and execution environment that runs on large clusters of commodity machines.

HDFS: A distributed filesystem that runs on large clusters of commodity machines.

Pig: A data flow language and execution environment for exploring very large datasets. Pig runs on HDFS and MapReduce clusters [6].

Hive: A distributed data warehouse. Hive manages data stored in HDFS and provides a query language based on SQL (and which is translated by the runtime engine to MapReduce jobs) for querying the data [7].

HBase: A distributed, column-oriented database. HBase uses HDFS for its underlying storage, and supports both batch-style computations using MapReduce and point queries (random reads) [18].

ZooKeeper: A distributed, highly available coordination service. ZooKeeper provides primitives such as distributed locks that can be used for building distributed applications [19].

Sqoop: A tool for efficiently moving data between relational databases and HDFS.

Cascading: MapReduce is very powerful as a general-purpose computing framework, but writing applications in the Hadoop Java API for MapReduce is daunting, due to the low-level abstractions of the API, the verbosity of Java, and the relative inflexibility of the MapReduce for expressing many common algorithms. Cascading is the most popular high-level Java API that hides many of the complexities of MapReduce programming behind more intuitive pipes and data flow abstractions.

Twitter Scalding: Cascading is known to provide close abstraction to the daunting Java API using pipes and custom data flows. It still suffers from limitations and verbosity of Java. Scalding is Scala API on top of cascading that indents to remove most of the boilerplate Java code and provides concise implementations of common data analytics and manipulation functions similar to SQL and Pig. Scalding also provides algebra and matrix models, which are useful in implementing machine learning and other linear algebra-dependent algorithms.

Cascalog: Cascalog is similar to Scalding in the way it hides the limitations of Java behind a powerful Clojure API for cascading. Cascalog includes logic programming constructs inspired by Datalog. The name is derived from Cascading + Datalog.

Impala: It is a scalable parallel database technology to Hadoop, which can be used to launch SQL queries on the data stored in HDFS and Apache HBase without any data movement or transformation. It is a massively parallel processing engine that runs natively on Hadoop.

Apache BigTop: It was originally a part of the Cloudera's CDH distribution, which is used to test the Hadoop ecosystem.

Apache Drill: It is an open-source version of Google Drell . It is used for interactive analysis on large-scale datasets. The primary goals of Apache Drill are real-time querying of large datasets and scaling to clusters bigger than 10,000 nodes. It is designed to support nested data, but also supports other data schemes like Avro and JSON. The primary language, *DrQL*, is SQL like [20].

Apache Flume: It is responsible data transfer between "source" and "sink", which can be scheduled or triggered upon an event. It is used to harvest, aggregate, and move large amounts of data in and out of Hadoop. Flume allows different data formats for sources, Avro, files, and sinks, HDFS and HBase. It also has a querying engine so that the user can transform any data before it is moved between sources and sinks.

Apache Mahout: it is a collection of scalable data mining and machine learning algorithms implemented in Java. Four main groups of algorithms are:

- Recommendations, a.k.a. collective filtering

- Classification, a.k.a categorization

- Clustering

- Frequent itemset mining, a.k.a parallel frequent pattern mining

It is not merely a collection of algorithms as many machine learning algorithms are non-scalable, the algorithms in Mahout are written to be distributed in nature and use the MapReduce paradigm for execution.

Oozie: it is used to manage and coordinate jobs executed on Hadoop.

2.3 Hadoop Distributed File System

The distributed file system in Hadoop is designed to run on commodity hardware. Although it has many similarities with other distributed file systems, the differences are significant. It is highly fault-tolerant and is designed to run on low-cost hardware. It also provides high-throughput to stored data, hence can be used to store and process large datasets. To enable this streaming of data it relaxes some POSIX standards. HDFS was originally built for the Apache Nutch project and later forked in to an individual project under Apache [21].

HDFS by design is able to provide reliable storage to large datasets, allowing high-bandwidth data streaming to user applications. By distributing the storage and computation across several servers, the resource can scale up and down with demand while remaining economical.

HDFS is different from other distributed file systems in the sense that it uses a write-once-read-many model that relaxes concurrency requirements, provides simple data coherency, and enables high-throughput data access [22]. HDFS prides on the principle and proves to be more efficient when the processing is done near the data rather than moving the data to the applications space. The data writes are restricted to one writer at a time. The bytes are appended to the end of the stream and are stored in the order written. HDFS has many notable goals:

- Ensuring fault tolerance by detecting faults and applying quick recovery methods.

- MapReduce streaming for data access.

- Simple and robust coherency model.

- Processing is moved to the data, rather than data to processing.

- Support heterogeneous commodity hardware and operating systems.

- Scalability in storing and processing large amounts of data.

- Distributing data and processing across clusters economically.

- Reliability by replicating data across the nodes and redeploying processing in the event of failures.

2.3.1 Characteristics of HDFS

Hardware Failure: Hardware failure is fairly common in clusters. A Hadoop cluster consists of thousands of machines, each of which stores a block of data. HDFS consists of a huge number of components and with that there is a good chance of failure among them at any point of time. The detection of these failures and the ability to quickly recover is part of the core architecture.

Streaming Data Access: Applications that run on the Hadoop HDFS need access to streaming data. These applications cannot be run on general-purpose file systems. HDFS is designed to enable large-scale batch processing, which is enabled by the high-throughput data access. Several POSIX requirements are relaxed to enable these special needs of high throughput rates.

Large Data Sets: The HDFS-based applications feed on large datasets. A typical file size is in the range of high gigabytes to low terabytes. It should provide high data bandwidth and support millions of files across hundreds of nodes in a single cluster.

Simple Coherency Model: The write-once-read-many access model of files enables high throughput data access as the data once written need not be changed, thus simplifying data coherency issues. A MapReduce-based application takes advantage of this model.

Moving compute instead of data: Any computation is efficient if it executes near the data because it avoids the network transfer bottleneck. Migrating the computation closer to the data is a cornerstone of HDFS-based programming. HDFS provides all the necessary application interfaces to move the computation close to the data prior to execution.

Heterogeneous hardware and software portability: HDFS is designed to run on commodity hardware, which hosts multiple platforms. This features helps widespread adoption of this platform for large-scale computations.

Drawbacks:

Low-latency data access: The high-throughput data access comes at the cost of latency. Latency sensitive applications are not suitable for HDFS. HBase has shown promise in handling low-latency applications along with large-scale data access.

Lots of small files: The metadata of the file system is stored in the namenode's memory(master node). The limit to the number of files in the file-system is dependent on the namenode memory. Typically, each file, directory, and block takes about 150 bytes. For example, if you had one million files, each taking one block, you would need at least 300 MB of memory. Storing millions of files seems possible, but hardware is incapable of accommodating billions of files [24].

HDFS Architecture

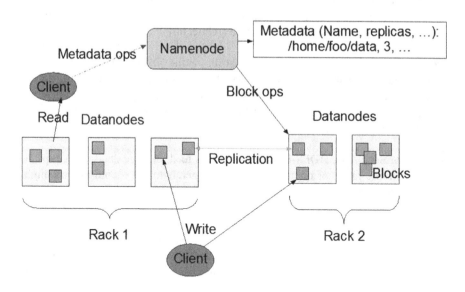

Fig. 2.2: HDFS Architecture [1]

2.3.2 Namenode and Datanode

The HDFS architecture is built in tandem with the popular master/slave architecture. An HDFS cluster consists of a master server that manages the file-system namespace and regulates files access, called the *Namenode*. Analogous to slaves, there are a number of *Datanodes*. These are typically one per cluster node and manage the data stored in the nodes. The HDFS file-system exists independently of the host file-system and allows data to be stored in its own namespace . The Namenode allows typical file-system operations like opening, closing, and renaming files and directories. It also maintains data block and Datanode mapping information. The Datanodes handle the read and write requests. Upon Namenode's instructions the Datanodes perform block creation, deletion, and replications operations. HDFS architecture is given in Figure 2.2

Namenode and Datanodes are software services provided by HDFS that are designed to run on heterogeneous commodity machines. These applications typically run on Unix/Linux-based operating systems. Java programming language is used to build these services. Any machine that supports the Java runtime environment can run Namenode and Datanode services. Given the highly portable nature of Java programming language HDFS can be deployed on wide range of machines. A typical cluster installation has one dedicated machine that acts as a master running the Namenode service. Other machines in the cluster run one instance of Datanode service per node. Although you can run multiple Datanodes on one machine, this practice is rare in real-world deployments.

A single instance of Namenode/master simplifies the architecture of the system. It acts as an arbitrator and a repository to all the HDFS meta-data. The system is designed such that there is data flow through the Namenode. Figure 2.3 shows how the Hadoop ecosystem interacts together.

2.3.3 File System

The file organization in HDFS is similar to the traditional hierarchical type. A user or an application can create and store files in directories. The namespace hierarchy is similar to other file-systems in the sense that one can create, remove, and move files from one directory to another, or even rename a file. HDFS does not support hard-links or soft-links.

Any changes to the file-system namespace is recorded by the Namenode. An application can specify the replication factor of a file that should be maintained by HDFS. The number of copies of a file is called the replication factor.

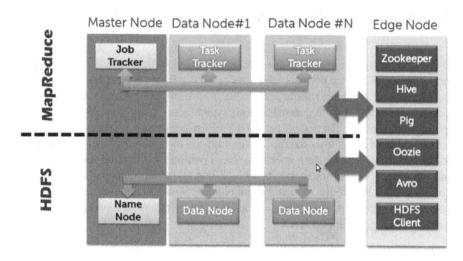

Fig. 2.3: Interaction between HDFS and MapReduce [1]

2.3.4 Data Replication

HDFS provides a reliable storage mechanism even though the cluster spans thousands of machines. Each file is stored as a sequence of blocks, where each block except the last one is of the same size. Fault tolerance is ensured by replicating the file blocks. The *block size* and *replication factor* is configurable for each file either by the user or an application. The replications factor can be set at file creation and can be modified later. Files in HDFS are write-once and strictly adhere to a single writer at a time property (Figure 2.4).

Namenode, acting as the master, takes all the decisions regarding data block replication. It receives a heartbeat, check Figure 2.5 on how it works, and block report from the Datanodes in the cluster. Heartbeat implies that the Datanode is functional and the block report provides a list of all the blocks in a Datanode.

Replication is vital to HDFS reliability and performance is improved by optimizing the placement of these replicas. This optimized placement contributes significantly to the performance and distinguishes itself from the rest of the file-systems. This feature requires much tuning and experience to get it right. Using a rack-aware replication policy improves data availability, reliability, and efficient network bandwidth utilization. This policy is the first of its kind and has seen much attention with better and sophisticated policies to follow.

Typically large HDFS instances span across multiple racks. Communication between these racks go through switches. Network bandwidth between machines in different racks is less than machines within the same rack.

Block Replication

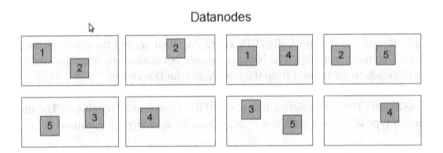

Namenode (Filename, numReplicas, block-ids, ...)
/users/sameerp/data/part-0, r:2, {1,3}, ...
/users/sameerp/data/part-1, r:3, {2,4,5}, ...

Datanodes

Fig. 2.4: Data Replication in Hadoop [22]

The Namenode is aware of the rack-ids each Datanode belongs to. A simple policy can be implemented by placing unique replicas in individual racks; in this way
even if an entire rack fails, although unlikely, the replica on another rack is still
available. However, this policy can be costly as the number of writes between racks
increases.

When the replication factor is, say 3, the placement policy in HDFS is put a replica
in one node per rack. This policy cuts the inter-rack write, which improves performance. The chance of rack failure is far less than node failure. This policy significantly reduces the network bandwidth used when reading data as a replica is placed
in only two unique racks instead of three. One-third of replicas are on one node and
one-third in another rack, while another third is evenly distributed across remaining racks. This policy improves performance along with data reliability and read
performance.

Replica Selection: Selecting a replica that is closer to the reader significantly
reduces the global bandwidth consumption and read latency. If a replica is present
on the same rack then it is chosen instead of another rack, similarly if the replica
spans data centers then the local data center hosting the replica is chosen.

Safe-mode: When HDFS services are started, the Namenode enters a safe-mode.
The Namenode receives heartbeats and block reports from the Datanodes. Each
block has a specified number of replicas and Namenode checks if these replicas
are present. It also checks with Datanodes through the heartbeats. Once a good per-

centage of blocks are present the Namenode exits the safe-mode (takes about 30s). It completes the remaining replication in other Datanodes.

2.3.5 Communication

The TCP/IP protocols are foundations to all the communication protocols built in HDFS. Remote Procedure Calls (RPCs) are designed around the client protocols and the Datanode protocols. The Namenode does not initiate any procedure calls, it only responds to the request from the clients and the Datanodes.

Robustness: The fault-tolerant nature of HDFS ensures data reliability. The three common type of failures are Namenode, Datanode, and network partitions.

Data Disk Failure, Heartbeats, and Re-replication: Namenode receives periodic heartbeats from the Datanodes. A network partition, failure of a switch, can cause all the Datanodes connected via that network to be invisible to the Namenode. Namenode used heartbeats to detect the condition of nodes in the cluster, it marks them dead if there are no recent heartbeats and does not forward any I/O requests. Any data part of the dead Datanode is not available and the Namenode keeps track of these blocks and triggers replications whenever necessary. There are several reasons to re-replicate a data block: Datanode failure, corrupted data block, storage disk failure, increased replication factor .

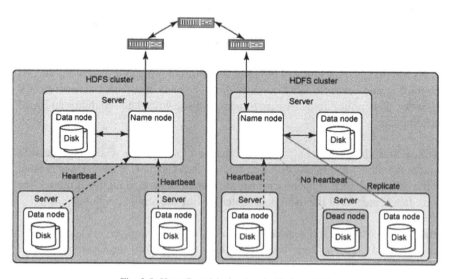

Fig. 2.5: Heart Beat Mechanism in Hadoop [22]

Cluster Rebalancing: Data re-balancing schemes are compatible with the HDFS architecture. It automatically moves the data from a Datanode when its free space falls below a certain threshold. In case a file is repeatedly used in an application, a scheme might create additional replicas of the file to re-balance the thirst for the data on the file in the cluster.

Data Integrity: Data corruption can occur for various reasons like storage device failure, network faults, buggy software, etc. To recognize corrupted data blocks HDFS implements a checksum to check on the contents of the retrieved HDFS files. Each block of data has an associated checksum that is stored in a separate hidden file in the same HDFS namespace. When a file is retrieved the quality of the file is checked with the checksum stored, if not then the client can ask for that data block from another Datanode (Figure 2.6).

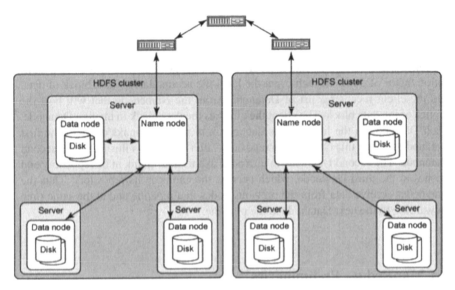

Fig. 2.6: Each cluster contains one Namenode. This design facilitates a simplified model for managing each Namespace and arbitrating data distribution [22]

2.3.6 Data Organization

Data Blocks: HDFS by design supports very large files. Applications written to work on HDFS write their data once and read many times, with reads at streaming speeds. A typical block size used by HDFS is 64 MB. Each file is chopped into 64 MB chunks and replicated.

Staging: A request to create a file does not reach the Namenode immediately. Initially, The HDFS client caches the file data into a temporary local file. Application writes are redirected to this temporary files. When the local file accumulates content over the block size, the client contacts the Namenode. The Namenode creates a file in the file-system and allocates a data block for it. The Namenode responds to the client with the Datanode and data block identities. The client flushes the block of data from the temporary file to this data block. If the file is closed the Namenode is informed, and it starts a file creation operation into a persistent store. Suppose, if the Namenode dies before commit the file is lost.

The benefits of the above approach is to allow streaming write to files. If a client writes a file directly to a remote file without buffering, the network speed and network congestion impacts the throughput considerably. There are earlier file-systems that successfully use client side caching to improve performance. In case of HDFS a few POSIX rules have been relaxed to achieve high performance data uploads.

Replication Pipelining: Local file caching mechanism described in the previous section is used for writing data to an HDFS file. Suppose the HDFS file has a replication factor of three. In such event the local file accumulates a data block of data, and the client receives a list of Datanodes from the Namenode that will host the replica of the data block. The client then flushes the data block in the first Datanode. Each Datanode in the list receives data block in smaller chunks of 4KB, the first Datanode persists this chunk in its repository after which it is flushed to the second Datanode. The second Datanode in turn persists this data chunk in its repository and flushes to the third Datanode, which persists the chunk in its repository. Thus the Datanodes receive data from the previous nodes in a *pipeline* and at the same time forwarding to the next Datanode in the pipeline.

2.4 MapReduce Preliminaries

Functional Programming: MapReduce framework facilitates computing with large volumes of data in parallel. This requires the workload to be divided across several machines. This model scales because there is no intrinsic sharing of data as the communication overhead needed to keep the data synchronized prevents the cluster from performing reliably and efficiently at large scale.

All the data elements of MapReduce cannot be updated, i.e., immutable. If you change the (key, value) pair in a mapping task, the change is not persisted. A new output (key, value) pair is created before it is forwarded by the Hadoop system into the next phase of execution.

MapReduce: Hadoop *MapReduce* is a simple software framework used to process vast amounts of data in the scale of higher terabytes in parallel on large clusters of commodity hardware in a reliable, fault-tolerant manner. A MapReduce job splits the dataset into independent chunks that are processed by many *map* tasks in parallel. The framework then sorts the output of this *map* which are then input to the *reduce* task. The input and output of the job are stored on the file-system. The framework also takes care of scheduling, monitoring, and re-launching failed tasks. Typically, in Hadoop the storage and compute nodes are the same, the MapReduce framework and HDFS run on the same set of machines. This allows for the framework to schedule jobs on nodes where data are already present, resulting in high-throughput across the cluster.

The framework consists of one *JobTracker* on the master/Namenode and one *Task-Tracker* per cluster node/Datanodes. The master is responsible for scheduling jobs, which are executed as tasks on the Datanodes. The Namenode is responsible for monitoring, re-executing failed tasks.

List Processing: MapReduce as paradigm transforms a list of input data into another set of output lists. The list processing idioms are *map* and *reduce*. These are principles of several functional programming languages like Scheme, LISP, etc.

Mapping Lists: A list of items provided one at a time to a function called the *mapper*, which transforms these elements one at a time to an output data element. For example, a function like *toUpper* converts a string into its uppercase version, and is applied to all the items in the list; the input string is not modified but a new transformed string is created and returned (Figure 2.7).

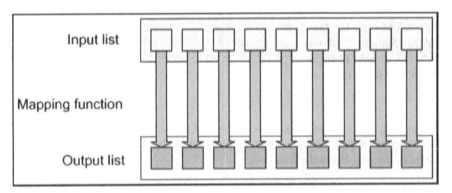

Fig. 2.7: Mapping creates a new output list by applying a function to individual elements of an input list [22].

Reducing Lists: The intermediate outputs of the mappers are sent to a *reducer* function that receives an iterator over input values. It combines these values returning a single output (Figure 2.8).

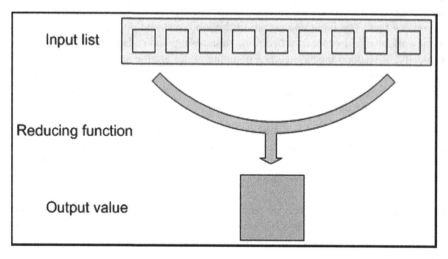

Fig. 2.8: reducing a list iterates over the input values to produce an aggregate value as output [22].

Keys and Values: : Every value in an input for MapReduce program has a key associated to it. Keys identify related values. A collection of timestamped speedometer readings, the value, from different cars has the license plate number, the key, associated to it.

AAA-123 65mph, 12:00pm
ZZZ-789 50mph, 12:02pm
AAA-123 40mph, 12:05pm
CCC-456 25mph, 12:15pm
...

The mapping and reducing functions receive not just values but (key, value) pairs. The output of each of these functions is the same: both a key and a value must be emitted to the next list in the data flow.

MapReduce is also less strict than other languages about how the Mapper and Reducer work. In more formal functional mapping and reducing settings, a mapper must produce exactly one output element for each input element, and a reducer must produce exactly one output element for each input list. In MapReduce, an arbitrary number of values can be output from each phase; a mapper may map one input into zero, one, or one hundred outputs. A reducer may compute over an input list and emit one or a dozen different outputs.

Keys divide the reduce space : A reducing function turns a large list of values into one (or a few) output values. In MapReduce, all of the output values are not usually reduced together. All of the values with the same key are presented to a single reducer together. This is performed independently of any reduce operations occurring on other lists of values, with different keys attached (Figure 2.9).

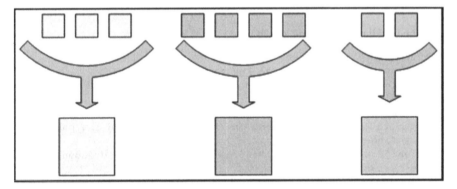

Fig. 2.9: Different colors represent different keys. All values with the same key are presented to a single reduce task [22].

2.5 Prerequisites for Installation

Hadoop cluster installation requires the following software packages:

1. Linux-based operating system, preferably Ubuntu

2. Java version 1.6 or higher

3. Secure Shell Access across the machines

Step 1: Check for Java in the systems:

```
user@ubuntu:~$ java -version
java version "1.6.0_27"
OpenJDK Runtime Environment (IcedTea6 1.12.6) (6b27-1.12.6-1
    ubuntu0.12.04.2)
OpenJDK 64-Bit Server VM (build 20.0-b12, mixed mode)
```

If the above command does not yield the correct output.

```
user@ubuntu:~$ sudo apt-get install openjdk-6-jdk
```

The installation is available in:

```
/usr/lib/jvm/java-6-openjdk-amd64
```

Step 2: The communication between nodes in the cluster happens via *ssh*. In a multi-node cluster setup of communication is between individual nodes, while in sing-node cluster, localhost acts as server.

Execute the following command to install ssh:

```
user@ubuntu:~$ sudo apt-get install openssh-server openssh-
    client
```

After the installation, generate an ssh key:

```
user@ubuntu:~$ ssh-keygen -t rsa -P ""
Generating public/private rsa key pair.
Enter file in which to save the key (/home/user/.ssh/id_rsa):
Created directory '/home/hduser/.ssh'.
Your identification has been saved in /home/user/.ssh/id_rsa.
Your public key has been saved in /home/user/.ssh/id_rsa.pub.
The key fingerprint is:
9b:82:ea:58:b4:e0:35:d7:ff:19:66:a6:ef:ae:0e:d2 hduser@ubuntu
The key's randomart image is:
[...snipp...]
user@ubuntu:~$
```

Note : If there is an error you must check your ssh installation, repeat Step 2.

Step 3: In order to establish password-less communication, the public key is trans-fered to the slave machines from the master, which will be discussed in the following sections. In the case of single-node cluster, the slave and master are in the same machine but communication is over the localhost server.

```
user@ubuntu:~$ cat $HOME/.ssh/id_rsa.pub >> $HOME/.ssh/
    authorized_keys
```

Step 4: Test the password-less connection by creating an ssh connection to local-host.

```
user@ubuntu:~$ ssh localhost
The authenticity of host 'localhost (::1)' can't be established.
RSA key fingerprint is d7:87:25:47:ae:02:00:eb:1d:75:4f:bb:44:f9
    :36:26.
Are you sure you want to continue connecting (yes/no)? yes
Warning: Permanently added 'localhost' (RSA) to the list of
    known hosts.
Linux ubuntu 2.6.32-22-generic #33-Ubuntu SMP Wed Apr 28
    13:27:30 UTC 2010 i686 GNU/Linux
Ubuntu 10.04 LTS
[...snipp...]
user@ubuntu:~$
```

If the connection should fail, these general tips might help:

- Enable debugging with `ssh -vvv localhost` and investigate the error in detail.

- Check the SSH server configuration in `/etc/ssh/sshd_config`, in particular the options `PubkeyAuthentication` (which should be set to yes) and `AllowUsers` (if this option is active, add the user user to it). If you made any changes to the SSH server configuration file, you can force a configuration reload with `sudo /etc/init.d/ssh reload`.

2.6 Single Node Cluster Installation

Step 1: Download and extract Hadoop source

There are several ways to install Hadoop on your system:

- Ubuntu Hadoop Package(.deb)

- Hadoop Source Code

- Third Party Hadoop Distributions (Cloudera, Hortonworks etc)

We are going to install hadoop from its source. Download Hadoop 1.0.3 from the link below:

```
user@ubuntu:~$ wget https://dl.dropboxusercontent.com/u
    /26579166/hadoop-1.0.3.tar.gz
```

Unpack the Hadoop source code (I have assumed your current working directory as home directory, you can pick your own directory)

```
user@ubuntu:~$ sudo tar xzf hadoop-1.0.3.tar.gz
user@ubuntu:~$ sudo mv hadoop-1.0.3 hadoop
```

Hadoop source is present in path: `/home/user/hadoop`

Step 2: Update the system `.bashrc` file found in the `/home` folder. Set the following environment variables for Hadoop use:

- Set HADOOP_HOME to the extracted folder, as shown in Step 1.

- Set JAVA_HOME to the installed Java path, shown in Step 1 of Section 2.5.

- Append HADOOP_HOME/bin to the system path, so that the executables are visible system-wide.

```
# Set Hadoop-related environment variables
export HADOOP_HOME=/home/user/hadoop
export JAVA_HOME=/usr/lib/jvm/java-6-openjdk-amd64
export PATH=$PATH:$HADOOP_HOME/bin
```

Step 3: Configuring Hadoop is simple, in the sense the changes have to be made in the following files found in /home/user/hadoop/conf:

1. hadoop-env.sh: Set the JAVA_HOME environmental variable, this is specific to Hadoop environment settings.

   Change:

   ```
   # The java implementation to use.   Required.
   # export JAVA_HOME=/usr/lib/j2sdk1.5-sun
   ```

   to:

   ```
   # The java implementation to use.   Required.
   export JAVA_HOME=/usr/lib/jvm/java-6-openjdk-amd64
   ```

2. core-site.xml: This file sets the HDFS path, where the file-system is installed and the data is stored. Create a folder /home/user/hadoop_tmp, which will be used as the storage location. Since the file is of xml format, the variables are set as property.

   Change:

   ```
   <property>
     <name>hadoop.tmp.dir</name>
     <value>Enter absolute path here</value>
     <description>A base for other temporary directories.
        </description>
   </property>
   ```

   To:

   ```
   <property>
     <name>hadoop.tmp.dir</name>
     <value>/home/user/hadoop_tmp</value>
     <description>A base for other temporary directories.
        </description>
   </property>
   ```

   Final core-site.xml looks like this:

   ```
   <?xml version="1.0"?>
   <?xml-stylesheet type="text/xsl" href="configuration.xsl"?>

   <configuration>
   <property>
     <name>hadoop.tmp.dir</name>
     <value>/home/user/hadoop_tmp</value>
     <description>A base for other temporary directories.
        </description>
   </property>
   ```

```
<property>
  <name>fs.default.name</name>
  <value>hdfs://localhost:54310</value>
  <description>The name of the default file system. A URI
      whose scheme and authority determine the FileSystem
      implementation. The uri's scheme determines the config
      property (fs.SCHEME.impl) naming the FileSystem
      implementation class.  The uri's authority is used to
      determine the host, port, etc. for a filesystem.
  </description>
</property>
</configuration>
```

3. `mapred-site.xml`: This file contains the configuration paramters set/unset for MapReduce jobs. One such parameter is the URL for `JobTracker`.

   Add the following code to the file:

```
<property>
  <name>mapred.job.tracker</name>
  <value>localhost:54311</value>
  <description>The host and port that the MapReduce job
      tracker runs at. If "local", then jobs are run in-
      process as a single map and reduce task.
  </description>
</property>
```

4. `hdfs-site.xml`: The replication factor is set/unset in the file by adding the following configuration setting:

```
<property>
  <name>dfs.replication</name>
  <value>1</value>
  <description>Default block replication. The actual number
      of replications can be specified when the file is
      created. The default is used if replication is not
      specified in create time.
  </description>
</property>
```

Step 4: Formatting file-system

So far the environmental variables and configuration files have been changed to suit the needs of a cluster. Installing the Hadoop Distributed File System(HDFS) means to format the installation path, as shown in Step 3, that is, `/home/user/hadoop_tmp`.

Execute the following command:

```
user@ubuntu:~$ ./hadoop/bin/hadoop namenode -format
```

The expected output of successful format looks like below:

```
user@ubuntu:~$ ./hadoop/bin/hadoop namenode -format
13/10/18 10:16:39 INFO namenode.NameNode: STARTUP_MSG:
/************************************************************
STARTUP_MSG: Starting NameNode
STARTUP_MSG:    host = ubuntu/127.0.1.1
STARTUP_MSG:    args = [-format]
STARTUP_MSG:    version = 1.0.3
STARTUP_MSG:    build = https://svn.apache.org/repos/asf/hadoop/
    common/branches/branch-1.0 -r 1335192; compiled by 'hortonfo'
     on Tue May  8 20:31:25 UTC 2012
************************************************************/
13/10/18 10:16:39 INFO util.GSet: VM type       = 64-bit
13/10/18 10:16:39 INFO util.GSet: 2% max memory = 17.77875 MB
13/10/18 10:16:39 INFO util.GSet: capacity      = 2^21 = 2097152
    entries
13/10/18 10:16:39 INFO util.GSet: recommended=2097152, actual
    =2097152
13/10/18 10:16:39 INFO namenode.FSNamesystem: fsOwner=user
13/10/18 10:16:39 INFO namenode.FSNamesystem: supergroup=
    supergroup
13/10/18 10:16:39 INFO namenode.FSNamesystem: isPermissionEnabled
    =true
13/10/18 10:16:39 INFO namenode.FSNamesystem: dfs.block.
    invalidate.limit=100
13/10/18 10:16:39 INFO namenode.FSNamesystem:
    isAccessTokenEnabled=false accessKeyUpdateInterval=0 min(s),
    accessTokenLifetime=0 min(s)
13/10/18 10:16:39 INFO namenode.NameNode: Caching file names
    occuring more than 10 times
13/10/18 10:16:39 INFO common.Storage: Image file of size 109
    saved in 0 seconds.
13/10/18 10:16:40 INFO common.Storage: Storage directory /home/
    user/hadoop_tmp/dfs/name has been successfully formatted.
13/10/18 10:16:40 INFO namenode.NameNode: SHUTDOWN_MSG:
/************************************************************
SHUTDOWN_MSG: Shutting down NameNode at ubuntu/127.0.1.1
************************************************************/
user@ubuntu:~$
```

Note: Always stop the cluster before formatting.

Step 5: After the file-system is installed, start the cluster to check if all the framework daemons are working.

Run the command:

```
user@ubuntu:~$ /usr/local/hadoop/bin/start-all.sh
```

This will startup a *Namenode*, *Datanode*, *Jobtracker* and a *Tasktracker* on your machine as daemon services.

```
user@ubuntu:~$ ./hadoop/bin/start-all.sh
starting namenode, logging to /home/user/hadoop/libexec/../logs/
    hadoop-user-namenode-ubuntu.out
localhost: starting datanode, logging to /home/user/hadoop/
    libexec/../logs/hadoop-user-datanode-ubuntu.out
localhost: starting secondarynamenode, logging to /home/user/
    hadoop/libexec/../logs/hadoop-user-secondarynamenode-ubuntu.
    out
starting jobtracker, logging to /home/user/hadoop/libexec/../
    logs/hadoop-user-jobtracker-ubuntu.out
localhost: starting tasktracker, logging to /home/user/hadoop/
    libexec/../logs/hadoop-user-tasktracker-ubuntu.out
user@ubuntu:~$
```

To check if all the daemon services are functional, use the built-in Java process tool:
jps

```
user@ubuntu:~$ jps
21128 Jps
20328 DataNode
20596 SecondaryNameNode
20689 JobTracker
20976 TaskTracker
19989 NameNode
```

Note: If any of above services are missing, debug by looking at the logs in
/home/user/hadoop/logs.

Step 6: Stopping the cluster

To stop all the daemons execute the following command:

```
user@ubuntu:~$ ./hadoop/bin/stop-all.sh
```

output:

```
user@ubuntu:~$ ./hadoop/bin/stop-all.sh
stopping jobtracker
localhost: stopping tasktracker
stopping namenode
localhost: stopping datanode
localhost: stopping secondarynamenode
user@ubuntu:~$
```

2.7 Multi-node Cluster Installation

To enable simple installation of a multi-node cluster, it is recommended that every node in the proposed cluster have Hadoop installed and configured as per the single-node installation instructions in Section 2.6. In the case of single node both master and slave are on the same machine, after installation it is only a matter of identifying the "master" and "slave" machines.

Here we try to create a multi-node cluster with two nodes, as shown in Figure 2.10. We will modify the Hadoop configuration to make one Ubuntu box the "master" (which will also act as a slave) and the other Ubuntu box a "slave".

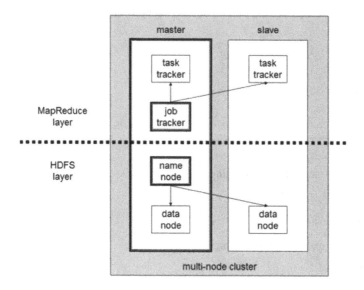

Fig. 2.10: Multi-node Cluster Overview

Note 1: We will call the designated master machine just the "master" from now on and the slave-only machine the "slave". We will also give the two machines these respective hostnames in their networking setup, most notably in /etc/hosts. If the hostnames of your machines are different (e.g. "node01") then you must adapt the settings in this section as appropriate.

Note 2: Shutdown each single-node cluster with bin/stop-all.sh before continuing if you haven't done so already.

Step 1: Networking

While creating a multi-node cluster it is assumed that all the nodes in the proposed cluster are reachable over the network. Ideally, these are connected to one hub or switch. For example, machines are connected to a switch and are identified by the address sequence `192.168.0.x/24`.

Since we have a two-node cluster, IP addresses `192.168.0.1` and `192.168.0.2` are assigned to nodes respectively. Update `/etc/hosts` on both machines with the following lines:

```
192.168.0.1     master
192.168.0.2     slave
```

Step 2: Secure Shell Access across the network.

In Sections 2.6 and 2.5 password-less logins are created by generating an ssh key and adding that key to `authorized` key listing. Similarly, the key generated on the `master` node is transfered to each of the `slave` nodes. The keys must be transfered so that Hadoop jobs can be started and stopped without the user authentication.

execute the following command from the `master` node.

```
user@master:~$ ssh-copy-id -i $HOME/.ssh/id_rsa.pub user@slave
```

This will prompt you to enter the login password for user `user` on the `slave` machine, copy the `master` public key and assign correct directory and permissions.

```
user@master:~$ ssh-copy-id -i $HOME/.ssh/id_rsa.pub user@slave
```

After successful transfer of the keys, test the connection to each of the `slave` nodes from the `master`. The step is also needed to save slaves host key fingerprint to the `hduser@masters known_hosts` file.

`master` to `master`:

```
user@master:~$ ssh master
The authenticity of host 'master (192.168.0.1)' can't be
    established.
RSA key fingerprint is 3b:21:b3:c0:21:5c:7c:54:2f:1e:2d:96:79:eb
    :7f:95.
Are you sure you want to continue connecting (yes/no)? yes
Warning: Permanently added 'master' (RSA) to the list of known
    hosts.
Linux master 2.6.20-16-386 #2 Fri Oct 18 20:16:13 UTC 2013 i686
...
user@master:~$
```

`master` to `slave`:

```
user@master:~$ ssh slave
The authenticity of host 'slave (192.168.0.2)' can't be
    established.
RSA key fingerprint is 74:d7:61:86:db:86:8f:31:90:9c:68:b0
    :13:88:52:72.
Are you sure you want to continue connecting (yes/no)? yes
Warning: Permanently added 'slave' (RSA) to the list of known
    hosts.
Ubuntu 12.04
...
user@slave:~$
```

Step 3: Multi-node configuration

With reference Step 3 in Section 2.6 the following files in the /home/user/ hadoop/conf directory need to be changed:

1. core-site.xml on all machines

   Change the fs.default.name parameter, which specifies the NameNode (the HDFS master) host and port. In our case, this is the master machine.

   Change:

```
<property>
  <name>fs.default.name</name>
  <value>hdfs://localhost:54310</value>
</property>
```

   To:

```
<property>
  <name>fs.default.name</name>
  <value>hdfs://master:54310</value>
</property>
```

2. hdfs-site.xml on all machines

   change the |dfs.replication— parameter (in |conf/hdfs-site.xml—) which specifies the default block replication. It defines how many machines a single file should be replicated to before it becomes available.

   The default value of dfs.replication is 3. However, we have only two nodes available, so we set dfs.replication to 2.

   Change:

```
<property>
  <name>dfs.replication</name>
  <value>1</value>
</property>
```

To:

```
<property>
  <name>dfs.replication</name>
  <value>2</value>
</property>
```

3. `mapred-site.xml` on all machines

   Change the `mapred.job.tracker` parameter which specifies the `Job Tracker` (MapReduce master) host and port. Again, this is the `master` in our case.

   Change:

```
<property>
  <name>mapred.job.tracker</name>
  <value>localhost:54311</value>
  </property>
```

   To:

```
<property>
  <name>mapred.job.tracker</name>
  <value>master:54311</value>
</property>
```

4. `masters` file, on the `master` node

   This file identifies the machines as the `Namenode` and `Secondary Namenodes` associated by the IP addresses. Enter the hostname that has been allocated as the `master` in the `/etc/hosts` file.

```
master
```

5. `slaves` file, on the `master` node

   This file lists the hosts, one per line, where the Hadoop slave daemons (DataNodes and TaskTrackers) will be run. Instead of wasting the compute resource of the master we add `master` as a slave so that it can share the load of cluster.

```
master
slave
```

   If you have additional slave nodes, just add them to the `slaves` file, one hostname per line.

```
master
slave
anotherslave01
anotherslave02
anotherslave03
```

Note: Typically, one machine in the cluster is designated as the NameNode and another machine the as JobTracker, exclusively. These are the actual "master nodes". The rest of the machines in the cluster act as both DataNode and TaskTracker. These are the slaves or "worker nodes".

Additional Configuration Parameters: There are some other configuration options worth studying

In file conf/mapred-site.xml:

- *mapred.local.dir*

  Determines where temporary MapReduce data is written. It also may be a list of directories.

- *mapred.map.tasks*

  As a rule of thumb, use 10x the number of slaves (i.e., number of TaskTrackers).

- *mapred.reduce.tasks*

  As a rule of thumb, use

  num_tasktrackers * num_reduce_slots_per_tasktracker * 0.99.

  If num_tasktrackers is small

  use (num_tasktrackers − 1) * num_reduce_slots_per_tasktracker.

Step 4: Starting multi-node cluster

Starting the cluster is performed in two steps.

- We begin with starting the HDFS daemons: the Namenode daemon is started on master, and DataNode daemons are started on all slaves (here: master and slave).

  Run the command bin/start-dfs.sh on the machine you want the (primary) NameNode to run on. This will bring up HDFS with the NameNode running on the machine you ran the previous command on, and DataNodes on the machines listed in the conf/slaves file.

```
user@master:~$ ./hadoop/bin/start-dfs.sh
starting namenode, logging to /usr/local/hadoop/bin/../logs/
    hadoop-hduser-namenode-master.out
slave: Ubuntu 10.04
slave: starting datanode, logging to /usr/local/hadoop/bin
    /../logs/hadoop-hduser-datanode-slave.out
master: starting datanode, logging to /usr/local/hadoop/bin
    /../logs/hadoop-hduser-datanode-master.out
```

```
master: starting secondarynamenode, logging to /usr/local/
    hadoop/bin/../logs/hadoop-hduser-secondarynamenode-master.
    out
user@master:~$
```

At this point, the following Java processes should run on master

```
user@master:~$ jps
16017 Jps
14799 NameNode
14880 DataNode
14977 SecondaryNameNode
hduser@master:~$
```

and the following Java processes should run on every slave node.

```
user@master:~$ jps
15183 DataNode
16284 Jps
hduser@master:~$
```

- Then we start the MapReduce daemons: the JobTracker is started on master, and TaskTracker daemons are started on all slaves (here: master and
  slave).

```
user@master:~$ ./hadoop/bin/start-mapred.sh
```

At this point, the following Java processes should run on master

```
user@master:~$ jps
16017 Jps
14799 NameNode
15686 TaskTracker
14880 DataNode
15596 JobTracker
14977 SecondaryNameNode
hduser@master:~$
```

and the following Java processes should run on every slave node.

```
user@master:~$ jps
15183 DataNode
15897 TaskTracker
16284 Jps
hduser@master:~$
```

Step 5: Stopping the cluster

Similar to Step 4, the cluster is stopped in two steps:

- Stop the MapReduce daemons: the JobTracker is stopped on `master`, and Task-Tracker daemons are stopped on all slaves (here: `master` and `slave`).

  Run the command `bin/stop-mapred.sh` on the `JobTracker` machine. This will shut down the MapReduce cluster by stopping the JobTracker daemon running on the machine you ran the previous command on, and `TaskTrackers` on the machines listed in the `conf/slaves` file.

  In our case, we will run bin/stop-mapred.sh on master:

  ```
  user@master:~$./hadoop/bin/stop-mapred.sh
  stopping jobtracker
  slave: Ubuntu 10.04
  master: stopping tasktracker
  slave: stopping tasktracker
  user@master:~$
  ```

  At this point, the following Java processes should run on `master`

  ```
  user@master:~$ jps
  16017 Jps
  14799 NameNode
  14880 DataNode
  14977 SecondaryNameNode
  hduser@master:~$
  ```

  and the following Java processes should run on `slave`

  ```
  user@master:~$ jps
  15183 DataNode
  16284 Jps
  hduser@master:~$
  ```

- Stop the HDFS daemons: the NameNode daemon is stopped on `master`, and DataNode daemons are stopped on all slaves (here: `master` and `slave`).

  Run the command `bin/stop-dfs.sh` on the NameNode machine. This will shut down HDFS by stopping the NameNode daemon running on the machine you ran the previous command on, and DataNodes on the machines listed in the `conf/slaves` file.

  In our case, we will run `bin/stop-dfs.sh` on `master`:

  ```
  user@master:~$./hadoop/bin/stop-dfs.sh
  stopping namenode
  slave: Ubuntu 10.04
  slave: stopping datanode
  master: stopping datanode
  master: stopping secondarynamenode
  user@master:~$
  ```

  At this point, only the following Java processes should run on `master`

```
user@master:~$ jps
18670 Jps
user@master:~$
```

and the following on `slave`.

```
user@slave:~$ jps
18894 Jps
user@slave:~$
```

2.8 Hadoop Programming

Running `wordcount` Hadoop MapReduce job. This program reads text files and counts how often words occur. The input is text files and the output is text files, each line of which contains a word and the count of how often it occurred, separated by a tab.

Download input data: The Notebooks of Leonardo Da Vinci

```
user@ubuntu:~$ wget http://www.gutenberg.org/cache/epub/5000/
    pg5000.txt
```

output

```
user@ubuntu:~$ wget http://www.gutenberg.org/cache/epub/5000/
    pg5000.txt
--2013-10-18 11:02:36--  http://www.gutenberg.org/cache/epub
    /5000/pg5000.txt
Resolving www.gutenberg.org (www.gutenberg.org)... 152.19.134.47
Connecting to www.gutenberg.org (www.gutenberg.org)
    |152.19.134.47|:80... connected.
HTTP request sent, awaiting response... 200 OK
Length: 1423803 (1.4M) [text/plain]
Saving to: 'pg5000.txt'

100%[===================================>] 14,23,803   241K/s
    in 5.8s

2013-10-18 11:02:43 (241 KB/s) - 'pg5000.txt' saved
    [1423803/1423803]
```

Restart the Hadoop cluster: using script `start-all.sh`

```
user@ubuntu:~$ ./hadoop/bin/start-all.sh
```

Copy local example data to HDFS: using `copyFromLocal` dfs option.

```
user@ubuntu:~$ ./hadoop/bin/hadoop dfs -copyFromLocal pg5000.txt
    input
user@ubuntu:~$ ./hadoop/bin/hadoop dfs -ls
Found 1 items
-rw-r--r--   1 user supergroup      1423803 2013-10-18 11:35 /user
    /user/input
```

Run wordcount program: The wordcount program has already been compiled and stored as a jar file. This jar file comes with the downloaded Hadoop source. The wordcount example can be found in the `hadoop-examples-1.0.3.jar`.

```
user@ubuntu ./hadoop/bin/hadoop jar hadoop-examples-1.0.3.jar
    wordcount input output
```

This command will read all the files in the HDFS directory `/user/user/input`, process it, and store the result in the HDFS directory `/user/user/output`.

```
user@ubuntu:~$ ./hadoop/bin/hadoop jar hadoop-examples-1.0.3.jar
    wordcount input output
13/10/18 13:47:34 INFO input.FileInputFormat: Total input paths
    to process : 1
13/10/18 13:47:34 INFO util.NativeCodeLoader: Loaded the native-
    hadoop library
13/10/18 13:47:34 WARN snappy.LoadSnappy: Snappy native library
    not loaded
13/10/18 13:47:34 INFO mapred.JobClient: Running job:
    job_201310181134_0001
13/10/18 13:47:35 INFO mapred.JobClient:  map 0% reduce 0%
13/10/18 13:47:48 INFO mapred.JobClient:  map 100% reduce 0%
13/10/18 13:48:00 INFO mapred.JobClient:  map 100% reduce 100%
13/10/18 13:48:05 INFO mapred.JobClient: Job complete:
    job_201310181134_0001
.
.
.
13/10/18 13:48:05 INFO mapred.JobClient:       Virtual memory
    (bytes) snapshot=4165484544
13/10/18 13:48:05 INFO mapred.JobClient:       Map output records
    =251352
```

Check if the result is successfully stored in HDFS directory `/user/user/output`

```
user@ubuntu:~$ ./hadoop/bin/hadoop dfs -ls output
Found 3 items
-rw-r--r--   1 user supergroup            0 2013-10-18 13:48 /user
    /user/output/_SUCCESS
drwxr-xr-x   - user supergroup            0 2013-10-18 13:47 /user
    /user/output/_logs
-rw-r--r--   1 user supergroup       337648 2013-10-18 13:47 /user
    /user/output/part-r-00000
```

Inspect if the output has been correctly populated.

```
user@ubuntu:~$ hadoop dfs -cat /user/mak/output/part-r-00000
"(Lo)cra"    1
"1490    1
"1498,"  1
"35"     1
"40,"    1
"AS-IS".    1
"A_  1
"Absoluti    1
"Alack!  1
"Alack!"    1
.
.
.
```

Hadoop Web Interfaces: The progress of the Hadoop job can be monitored using the built-in web interfaces. These interfaces provide a graphical representation of the jobs.

http://localhost:50070/ — web UI of the NameNode daemon
http://localhost:50030/ — web UI of the JobTracker daemon
http://localhost:50060/ — web UI of the TaskTracker daemon

NameNode Web Interface (HDFS layer): It shows details of the storage capacity and usage, number of live nodes and dead nodes, access to logs, browse file-system, etc (Figure 2.11).

JobTracker Web Interface (MapReduce layer): It provides information about job statistics, job statuses if running/completed/failed and also job history logs (Figure 2.12).

TaskTracker Web Interface (MapReduce layer): It shows the status of running and non-running tasks. It also gives access to the tasktracker log files of that task.

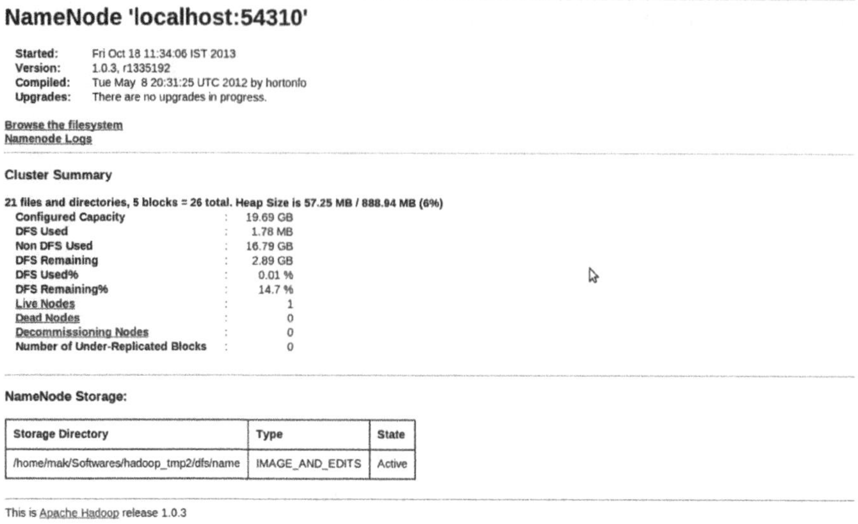

NameNode 'localhost:54310'

Started: Fri Oct 18 11:34:06 IST 2013
Version: 1.0.3, r1335192
Compiled: Tue May 8 20:31:25 UTC 2012 by hortonfo
Upgrades: There are no upgrades in progress.

Browse the filesystem
Namenode Logs

Cluster Summary

21 files and directories, 5 blocks = 26 total. Heap Size is 57.25 MB / 888.94 MB (6%)

Configured Capacity	:	19.69 GB
DFS Used	:	1.78 MB
Non DFS Used	:	16.79 GB
DFS Remaining	:	2.89 GB
DFS Used%	:	0.01 %
DFS Remaining%	:	14.7 %
Live Nodes	:	1
Dead Nodes	:	0
Decommissioning Nodes	:	0
Number of Under-Replicated Blocks	:	0

NameNode Storage:

Storage Directory	Type	State
/home/mak/Softwares/hadoop_tmp2/dfs/name	IMAGE_AND_EDITS	Active

This is Apache Hadoop release 1.0.3

Fig. 2.11: NameNode Web Interface (HDFS layer)

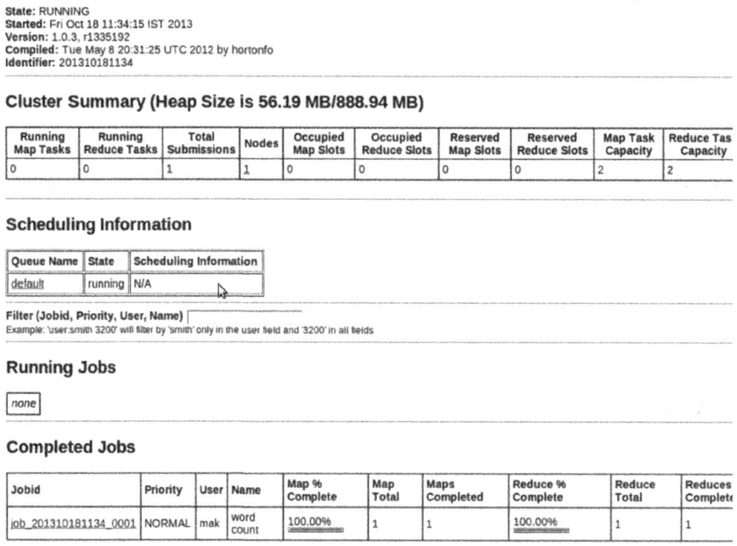

State: RUNNING
Started: Fri Oct 18 11:34:15 IST 2013
Version: 1.0.3, r1335192
Compiled: Tue May 8 20:31:25 UTC 2012 by hortonfo
Identifier: 201310181134

Cluster Summary (Heap Size is 56.19 MB/888.94 MB)

Running Map Tasks	Running Reduce Tasks	Total Submissions	Nodes	Occupied Map Slots	Occupied Reduce Slots	Reserved Map Slots	Reserved Reduce Slots	Map Task Capacity	Reduce Tas Capacity
0	0	1	1	0	0	0	0	2	2

Scheduling Information

Queue Name	State	Scheduling Information
default	running	N/A

Filter (Jobid, Priority, User, Name)
Example: 'user:smith 3200' will filter by 'smith' only in the user field and '3200' in all fields

Running Jobs

none

Completed Jobs

Jobid	Priority	User	Name	Map % Complete	Map Total	Maps Completed	Reduce % Complete	Reduce Total	Reduces Complete
job_201310181134_0001	NORMAL	mak	word count	100.00%	1	1	100.00%	1	1

Fig. 2.12: JobTracker Web Interface (MapReduce layer)

tracker_HackStation:localhost/127.0.0.1:56654 Task Tracker Status

Fig. 2.13: TaskTracker Web Interface (MapReduce layer).

2.9 Hadoop Streaming

Hadoop streaming is a utility that helps users to write and run MapReduce jobs using any executable or script as a mapper and/or reducer. Streaming is similar to a `pipe` operation in Linux. The text input is printed on the `stdin` stream, then passed to the Mapper script reading from `stdin`. The resulting output is written to the `stdout` stream which is then passed to Reducer. Finally, the reducer writes the output to a storage location, typically an HDFS directory.

The command line interface looks as follows.

```
$HADOOP_HOME/bin/hadoop jar \
    $HADOOP_HOME/hadoop-streaming.jar \
    -input myInputDirs \
    -output myOutputDir \
    -mapper /bin/cat \
    -reducer /bin/wc
```

Here the `-mapper` and `-reducer` are Linux executables that read and write to `stdin` and `stdout` respectively. The `hadoop-streaming.jar` utility creates a MapReduce job on the cluster and monitors its progress.

When the mapper tasks are executed, the input is converted into lines and fed into the stdin of the mapper task. The mapper processes these lines as they come and converts them to (key, value) pairs, which are then sent out to the stdout. These (key, value) pairs are then fed into the stdin of the reducer task launched after the completion of the mapper. The (key, value) pairs are reduced by the key and operation is performed on the values. The results are then pushed to the stdout that is written to persistent storage. This is how streaming works with the MapReduce framework.

Consequently, this is not limited to scripting languages. Java classes can also be plugged in as mapper classes.

```
$HADOOP_HOME/bin/hadoop  jar $HADOOP_HOME/hadoop-
   streaming.jar \
     -input myInputDirs \
     -output myOutputDir \
     -mapper org.apache.hadoop.mapred.lib.IdentityMapper \
     -reducer /bin/wc
```

Note: Streaming tasks exiting with nonzero status are considered failed.

Streaming Command Line Interface: Streaming supports streaming command options as well as generic command options. The general command line syntax is shown below.

Note: Be sure to place the generic options before the streaming options, otherwise the command will fail.

```
bin/hadoop command [genericOptions] [streamingOptions]
```

The Hadoop generic command options you can use with streaming are listed here [28]:

Commands	Description
-input directoryname or filename	(Required) Input location for mapper
-output directoryname	(Required) Output location for reducer
-mapper JavaClassName	(Required) Mapper executable
-reducer JavaClassName	(Required) Reducer executable
-file filename	Make the mapper, reducer, or combiner executable available locally on the compute nodes
-inputformat JavaClassName	Class you supply should return key/value pairs of Text class. If not specified, TextInputFormat is used as the default
-outputformat JavaClassName	Class you supply should take key/value pairs of Text class. If not specified, TextOutputformat is used as the default
-partitioner JavaClassName	Class that determines which reduce a key is sent to
-combiner JavaClassName	Combiner executable for map output

Any executable can be specified as a mapper or a reducer, if the file is not present in the nodes then −file option is used to tell Hadoop to package this file as part of the job and send across the cluster nodes.

```
$HADOOP_HOME/bin/hadoop   jar $HADOOP_HOME/hadoop-streaming.jar \
    -input myInputDirs \
    -output myOutputDir \
    -mapper myPythonScript.py \
    -reducer /bin/wc \
    -file myPythonScript.py
```

In addition to executables, other files like configuration files, dictionaries, etc., can also be shipped to the nodes while job submission using −file option.

```
$HADOOP_HOME/bin/hadoop   jar $HADOOP_HOME/hadoop-streaming.jar \
    -input myInputDirs \
    -output myOutputDir \
    -mapper myPythonScript.py \
    -reducer /bin/wc \
    -file myPythonScript.py \
    -file myDictionary.txt
```

Hadoop Streaming with Python: Python-based mapper and reducer are written as executables. They will read from and write to sys.stdin and sys.stdout respectively, similar to the discussion above. Here a simple wordcount program is implemented in python as follows:

Step 1: Map Step

The mapper executable reads the lines from sys.stdin, splits them into words, and prints out each word as a key and associated count 1 as the value. This file is stored as /home/user/mapper.py.

Note: Make sure the file has execution permission chmod +x /home/hduser/ mapper.py

```
#!/usr/bin/env python

import sys

for line in sys.stdin:
    line = line.strip()
    words = line.split()
    for word in words:
        print word, "1"
```

Step 2: Reduce step

The output of the mapper is passed as the input to the reducer. The reducer groups the words as keys and sums the value associated with each word, thus counting the frequency of words in the input file. This file is stored as `/home/user/reducer.py`.

Make sure the file has execution permission

```
chmod +x /home/user/reducer.py.
```

```python
#!/usr/bin/env python

from operator import itemgetter
import sys

current_word = None
current_count = 0
word = None

for line in sys.stdin:
    line = line.strip()

    word, count = line.split()

    try:
        count = int(count)
    except ValueError:
        continue

    if current_word == word:
        current_count += count
    else:
        if current_word:
            print current_word, current_count
        current_count = count
        current_word = word

if current_word == word:
    print current_word, current_count
```

Note: The input file must be transfered to the HDFS before executing the streaming job.

```
hduser@ubuntu:/usr/local/hadoop$ bin/hadoop jar contrib/
    streaming/hadoop-*streaming*.jar \
 -file /home/user/mapper.py \
 -mapper /home/user/mapper.py \
 -file /home/user/reducer.py  \
 -reducer /home/user/reducer.py \
 -input input -output output
```

If you want to modify some Hadoop settings on the fly like increasing the number of Reduce tasks, you can use the −D option:

```
hduser@ubuntu:/usr/local/hadoop$ bin/hadoop jar contrib/
    streaming/hadoop-*streaming*.jar -D mapred.reduce.tasks=16
    ...
```

References

1. Tom White, 2012, Hadoop: The Definitive Guide, O'reilly
2. Hadoop Tutorial, Yahoo Developer Network, http://developer.yahoo.com/hadoop/tutorial
3. Mike Cafarella and Doug Cutting, April 2004, Building Nutch: Open Source Search, ACM Queue, http://queue.acm.org/detail.cfm?id=988408.
4. Sanjay Ghemawat, Howard Gobioff, and Shun-Tak Leun, g, October 2003, The Google File System, http://labs.google.com/papers/gfs.html.
5. Jeffrey Dean and Sanjay Ghemawat, December 2004, MapReduce: Simplified Data Processing on Large Clusters, http://labs.google.com/papers/mapreduce.html
6. Yahoo! Launches World?s Largest Hadoop Production Application, 19 February 2008, http://developer.yahoo.net/blogs/hadoop/2008/02/yahoo-worlds-largest-production-hadoop.html.
7. Derek Gottfrid, 1 November 2007, Self-service, Prorated Super Computing Fun!, http://open.blogs.nytimes.com/2007/11/01/self-service-prorated-super-computing-fun/.
8. Google, 21 November 2008, Sorting 1PB with MapReduce, http://googleblog.blogspot.com/2008/11/sorting-1pb-with-mapreduce.html.
9. From Gantz et al., March 2008, The Diverse and Exploding Digital Universe, http://www.emc.com/collateral/analyst-reports/diverse-exploding-digital-universe.pdf
10. http://www.intelligententerprise.com/showArticle.jhtml?articleID=207800705, http://mashable.com/2008/10/15/facebook-10-billion-photos/, http://blog.familytreemagazine.com/insider/Inside+Ancestrycoms+TopSecret+Data+Center.aspx, and http://www.archive.org/about/faqs.php, http://www.interactions.org/cms/?pid=1027032.
11. David J. DeWitt and Michael Stonebraker, In January 2007 ?MapReduce: A major step backwards? http://databasecolumn.vertica.com/database-innovation/mapreduce-a-major-step-backwards
12. Jim Gray, March 2003, Distributed Computing Economics, http://research.microsoft.com/apps/pubs/default.aspx?id=70001
13. Apache Mahout, http://mahout.apache.org/
14. Think Big Analytics, http://thinkbiganalytics.com/leading_ big_ data_ dtechnologies/hadoop/
15. Jeffrey Dean and Sanjay Ghemawat, 2004, MapReduce: Simplified Data Processing on Large Clusters. Proc. Sixth Symposium on Operating System Design and Implementation.
16. Olston, Christopher, et al. "Pig latin: a not-so-foreign language for data processing." Proceedings of the 2008 ACM SIGMOD international conference on Management of data. ACM, 2008.
17. Thusoo, Ashish, et al. "Hive: a warehousing solution over a map-reduce framework." Proceedings of the VLDB Endowment 2.2 (2009): 1626-1629.
18. George, Lars. HBase: the definitive guide. " O'Reilly Media, Inc.", 2011.
19. Hunt, Patrick, et al. "ZooKeeper: Wait-free Coordination for Internet-scale Systems." USENIX Annual Technical Conference. Vol. 8. 2010.
20. Hausenblas, Michael, and Jacques Nadeau. "Apache drill: interactive Ad-Hoc analysis at scale." Big Data 1.2 (2013): 100-104.
21. Borthakur, Dhruba. "HDFS architecture guide." HADOOP APACHE PROJECT http://hadoop.apache.org/common/docs/current/hdfs design. pdf (2008).
22. [Online] IBM DeveloperWorks, http://www.ibm.com/developerworks/library/waintrohdfs/
23. Konstantin Shvachko, Hairong Kuang, Sanjay Radia, and Robert Chansler, May 2010, The Hadoop Distributed File System, Proceedings of MSST2010, http://storageconference.org/2010/Papers/MSST/Shvachko.pdf
24. [Online] Konstantin V. Shvachko, April 2010, HDFS Scalability: The limits to growth, pp. 6?16 http://www.usenix.org/publications/login/2010-04/openpdfs/shvachko.pdf
25. [Online] Micheal Noll, Single Node Cluster, http://www.michael-noll.com/tutorials/running-hadoop-on-ubuntu-linux-single-node-cluster/
26. [Online] Micheal Noll, Multi Node Cluster, http://www.michaelnoll.com/tutorials/running-hadoop-on-ubuntu-linux-multi-node-cluster/

27. [Online]   Micheal   Noll,   Hadoop   Streaming:Python,   http://www.michael-noll.com/tutorials/writing-an-hadoop-mapreduce-program-in-python/
28. Hadoop, Apache. "Apache Hadoop." 2012-03-07]. http://hadoop.apache.org (2011).

Chapter 3
Getting Started with Spark

Cluster computing has seen a rise in improved and popular computing models, in which clusters execute data-parallel computations on unreliable machines. This is enabled by software systems that provide locality-aware scheduling, fault tolerance, and load balancing. MapReduce [1] has become the front runner in pioneering this model, while systems like Map-Reduce-Merge [2] and Dryad [3] have generalized different data flow types. These systems are scalable and fault tolerant because they provide a programming model that enables users in creating acyclic data flow graphs to pass input data through a set of operations. This model enables the system to schedule and react to faults better without any user intervention. While this model can be applied to a lot applications, there are problems that cannot be solved efficiently by acyclic data flows.

3.1 Overview

Spark is a fault-tolerant and distributed data analytics tool capable of implementing large-scale data intensive applications on commodity hardware. Hadoop and other technologies have already popularized acyclic data flow techniques for building data intensive applications on commodity clusters, but these are not suitable for applications that reuse a working dataset for multiple parallel operations. Some of these applications are iterative machine learning algorithms and interactive data analysis tools. Spark addresses these problems, and is also scalable and fault-tolerant. In-order to accommodate these goals, Spark introduces a data storage and processing abstraction called Resilient Distributed Datasets (RDDs). An RDD is a read-only distribution of objects that are partitioned across a set of machines and can be rebuilt if a partition is lost [6].

Spark performs well in these cases, where Hadoop users have reported deficiency with MapReduce:

© Springer International Publishing Switzerland 2015
K.G. Srinivasa and A.K. Muppalla, *Guide to High Performance Distributed Computing*, Computer Communications and Networks, DOI 10.1007/978-3-319-13497-0_3

- **Iterative jobs**: Gradient-Descent is an excellent example of an algorithm that is repeatedly applied to the same dataset to optimize a parameter. While it is easy to represent each iteration as a MapReduce job, the data from each iteration has to be loaded from the disk, incurring a significant performance penalty.

- **Interactive analytics**: Interfaces like Pig [6] and Hive [7] [7] are commonly used running SQL queries on large datasets using Hadoop. Ideally this dataset is loaded into memory and queried repeatedly, but with Hadoop every query is executed a MapReduce job that incurs significant latency from disk read.

A mechanism by which processes can access shared data without inter-process communication is called Distributed Shared Memory (DSM). There are many approaches to achieving distributed memory, two of these approaches are discussed below:

- **Shared Virtual Memory (SVM)** : This concept is similar to paged virtual memory. The concept is to group all the distributed memories into a single address space. However, it does not take into account type of shared data, that is the page size is arbitrarily fixed irrespective of the type and size of shared data. It does not allow the programmer to define the type [8].

- **Object Distributed Shared Memory (ODSM)** : In this case objects, with access functions, are shared. Here, the programmer is allowed to define the type of data shared. Objects creation, access, and modification are controlled by the DSM manager. While SVM works at the operating systems level, ODSM proposes an alternative message passing programming model [6].

The challenges of implementing a DSM system includes addressing problems like data location, data access, sharing and locking data, and data coherence, etc. These problems have connections with transactional models, data migrations, concurrent programming, distributed systems etc.

The RDDs are an abstraction for distributed shared memory [9]. However, there are differences between RDDs and DSMs. First, RDDs provides a much restricted programming model but to enable data regeneration in the even of node failure. DSM system achieve fault tolerance by check-pointing, [10] while Spark regenerates lost partitions by using a lineage information from the RDD [11]. This means only the lost partitions are recomputed in parallel on other nodes, hence there is no need to revert to a previous checkpoint and risk losing computation progress. Also, there is no overhead if a node fails. Second, similar to MapReduce Spark pushes the computation to where data resides, instead of providing access to globally shared address space. There are other examples which have a restricted DSM programming model to improve reliability and performance. Munin [12] allows the programmers to annotate the variables so that there is a defined access protocol to access them. Linda implements a fault-tolerant tuple space programming model [13]. Thor provides an interface to persistent shared objects [14].

MapReduce model is suitable for Spark development, however these jobs operate on RDDs that can persist across several operations [5]. A MapReduce framework, Twister, allowed for storing static data in memory from long-lived map tasks [16]. Twister, however, does not implement fault tolerance. Spark's RDD is fault tolerant and generic compared to iterative MapReduce. A Spark program can alternate between running multiple operations across RDDs. Twister can only execute one map and one reduce function at a time. A user can define multiple datasets in Spark and perform data analytics. Broadcast variables in Spark functions similar to Hadoop's distribute cache which shares a file across all nodes in a cluster. However, broadcast variables can be reused across parallel operations.

3.2 Spark Internals

Resilient distributed datasets are an important abstraction in Spark that allow read-only collection of objects capable of rebuilding lost partitions across clusters. These RDDs can be reused in multiple parallel operations through memory caching. RDDs use lineage information about the lost partition in the rebuilding process.

Scala programming language [1] is used to implement Spark. It is a statically typed and high-level programming language running over the Java Virtual Machine (JVM) exposing functional programming constructs. The Scala interpreter can be modified such that Spark can be used interactively for defining RDDs, functions, variables etc. Spark is perhaps one of the first frameworks to allow an efficient general-purpose programming language to define interactive processes on large datasets (Figure 3.1).

The applications in Spark are called drivers. These contain operations that can be executed on a single-node or multi-node cluster. Spark can co-exist wit Hadoop by using a resource manager like Mesos [19].

Resilient Distributed Datasets

Systems like MapReduce [1] and Dryad [3] provides locality-aware scheduling, fault tolerance and load balancing features that simplify distributed programming allowing users to analyze large datasets using commodity clusters. However, Resilient Distributed Datasets (RDDs) is a distributed memory abstraction that allows the users to write in-memory computations, while still maintaining the advantages of current data flow systems like MapReduce. RDD performs exceptionally well on iterative algorithms, machine learning, and interactive data mining while the other data flows become inadequate. This is possible because RDDs provide only a read-only datasets avoiding the need to checkpoint which is common in other shared memory techniques.

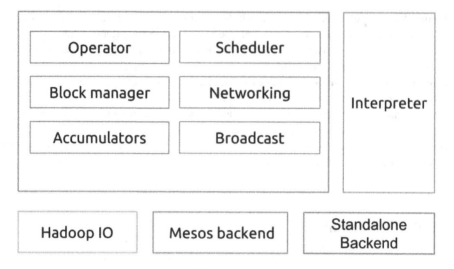

Fig. 3.1: Spark Architecture Overview [5]

Tools like Googles Pregel [15] allows for iterative graph processing algorithms, Twister [16] and HaLoop [17] provide an iterative MapReduce model. However, the communications are restricted. In contrast, RDDs allow users to generate intermediate results, partition control, and use operations of their choice rather than a set of steps like in MapReduce.

RDDs are easy to program and capable of efficiently expressing computations. Fault tolerance is perhaps most difficult to support. Check-pointing datasets has a huge disadvantage as it would require replicating datasets across clusters. This can slow down the machines due to bandwidth restrictions and memory storage restrictions. RDDs ensure fault tolerance by supporting stages of transformation. At any point of time, a partition can be recovered by repeating the transformation steps on the parent (Lineage). RDDs are well suited for applications that require executing a single function on many data records like graph and machine learning algorithms.

The contents of an RDD does not exist in physical storage, instead, a pointer to an RDD contains enough information to compute the RDD starting from data in storage. This allows for the reconstruction of an RDD. Since Spark is implemented using Scala, every RDD is a Scala object. RDDs can be created and transformed by:

- *file* in a file-system like HDFS.

- A Scala collection (e.g. array, list) can be *parallelized* in the Spark driver program dividing it into partitions and sent to multiple nodes.

- Using Spark transformations like *flatMap* [1] an existing RDD can be transformed from type A to type B.

- RDDs are lazy, that is, the partitions are created on demand when they are used in operations. However, a user can change its persistence by the following save:

  - The `cache` action indicates that after the initial computation of the RDD it is kept in memory for reuse. If there is not enough memory to cache the partitions, Spark is smart enough to recomute them when they are needed. This ensures the Spark program to continue execution. Spark addresses the trade-off between storing RDDs, speed of accessing it, probability of losing it, and cost of re-computing.

  - The `save` action write the contents of RDD to storage system like HDFS. The saved version is used in future operations on it.

RDDs Versus Distributed Shared Memory

RDDs are compared against DSMs to further the understanding of RDD capabilities. In DSMs, read and write operations are on the global address space, including traditional shared memory systems and shared distributed hash tables [21]. The generality of DSMs makes fault tolerance on commodity clusters difficult to implement. RDDs are created through bulk transformations while DSMs allow read and writes to each memory location. Although, RDDs are restricted to applications that perform bulk writes, it also allows for efficient fault tolerance. For example, RDDs avoid check-pointing using lineage. RDDs also recover only the partition without rolling back the whole system. RDDs also accommodate slow machines in the cluster by running backup tasks on them. Backup tasks are inefficient with DSMs as they make reads/writes to the same memory address. RDDs also provide run time efficiency by running tasks that are near to the data. Cache-based RDDs degrade gracefully when there is little or no memory. RDDs perform better through bulk reads on data, but they can also provide fine reads through key lookups on hash. A comparative understanding is provided in Table 3.1.

---

[1] flatMap has the same semantics as the map in MapReduce, but map is usually used to refer to a one-to-one function of type A $\Rightarrow$ B in Scala.

Table 3.1: Comparison of RDDs with DSMs [21]

Aspect	RDDs	Distributed Shared Memory
Reads	Bulk or fine-grained	Fine-grained
Writes	Bulk transformations	Fine-grained
Consistency	Trivial (immutable)	Up to app / runtime
Fault recovery	Fine-grained and low-overhead using lineage	Requires checkpoints and program rollback
Straggler mitigation	Possible using backup tasks	Difficult
Work placement	Automatic based on data locality	Up to app (runtimes aim for transparency)
Behavior if not enough RAM	Similar to existing data flow systems	Poor performance (swapping?)

Shared Variables

Spark lets users to create two types of shared variables; *Broadcast* variables and *Accumulators*. Programmers pass closures to Spark by invoking operations like *map*, *filter*, and *reduce*. These closures use the shared variables for computation. These variables are copied to worker nodes where the closures are executed [6].

- **Broadcast variables**: data representation like a lookup table is distributed once across all the nodes instead of packaging it with every closure. The Broadcast variables are used to wrap the value and are transfered to all the worker nodes [5].

- **Accumulators**: These are commonly used for *add-only* type of operations. They can be used to implement counters as in MapReduce and for parallel sums. Accumulators can be defined for any type that have an *add* operation with a *zero* value, which makes it is easy for fault tolerance [5].

When a broadcast variable b is created with a value v, it is stored in a shared file system. The serialized for b is a path to this file. When b is queried Spark first checks the cache, if not, it checks the file system. Each accumulator is identified by an unique ID. When an accumulator is saved, its ID and the *zero* value is stored in its serialized form. It is reset to zero when a task begins, and a copy if it is created for each thread that runs a task using thread-local variables. After each task, the worker informs the driver with the updates on accumulators. This driver updates the partitions after combining.

Working: Consider an example with a dataset containing error messages in a log file. These messages are stored in cache cachedErrs. We count its elements using map and reduce operations. Consider following set of operations:

```
val file = spark.textFile("hdfs://...")
val errs = file.filter(_.contains("ERROR"))
val cachedErrs = errs.cache()
val ones = cachedErrs.map(_ => 1)
val count = ones.reduce(_+_)
```

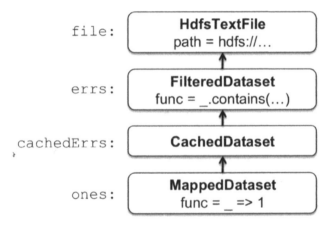

Fig. 3.2: Lineage chain for the distributed dataset objects [11]

The lineages for each RDD is captured for these chain of objects as shown in Figure 3.2. Each object contains a handle to its parent and information on how it was transformed [6]. Each RDD object implements the following three operations:

- getPartitions, which returns a list of partition IDs.

- getIterator(partition), which iterates over a partition.

- getPreferredLocations(partition), which is used for task scheduling to achieve data locality.

Tasks are created by Spark to operate on each partition of the dataset. These tasks are sent to the worked nodes as the partitions are distributed. A technique called delayed scheduling is used when sending tasks to partitions [20]. During the execution of a task, it calls getIterator API to start reading from the partition on that machine. RDDs differ only with the implementation of its communicating interface. For example, with HdfsTextFile interface the partitions are block IDs in HDFS and locations are block locations. getIterator opens a stream to read the data block. In a MappedDataset interface, the partitions and locations are the same as that of the parent, but map operations are applied on every element of the parent. In a CachedDataset interface, the getIterator API looks for a locally stored copy of partition and locations for each partition is similar to that of the parents. These locations are updated after the partition is cached on some node. This design helps fault detection, if a node fails the partitions for that node are rebuilt from the parent [5].

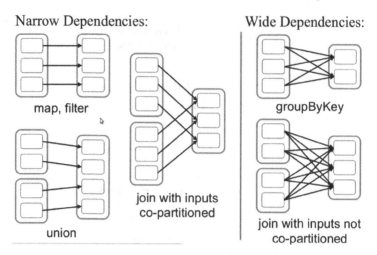

Fig. 3.3: Examples of narrow and wide dependencies. Each box is an RDD, with partitions shown as shaded rectangles [5].

Tasks are sent to the nodes by sending closures used to define the RDD and closures for operations like *reduce*. Java serialization is used to serialize Scala closures as Java objects and seamlessly executed on other machines.

Spark was built to accommodate several RDD transformations without modifying the scheduler each time. For every transformation, its lineage, function for computing the RDD from its parents, was captured. RDD is a collection of partitions which form the atomic pieces of the dataset. While developing this interface inter RDD dependencies needed to be addressed. They are a) *narrow dependencies*, where each child partition depends on the parent partition, e.g. a map task. b) *wide partitions*, where each child partition depends on the data from all the parent partitions, e.g. join task (Figure 3.3).

Job Scheduling

Spark scheduler depends on the RDD structure to create an efficient execution plan. The runJob function in Spark takes an RDD and a function as arguments. This interface can be used to express all the actions like count, collect, save, etc. The lineage graph captured while creating an RDD is used by the scheduler to create a DAG (Directed Acyclic Graph) of stages to execute as shown in Figure 3.4. For computing missing partitions as a fault-tolerant mechanism, a task is launched at each stage after the parents end. In-order to minimize communication the tasks are launched based on data locality. In the case of narrow dependency measures, the locations are given to the scheduler and the tasks are redirected, with wide dependencies intermediate records on the nodes are regenerated from the parents. Scheduler also performs various lookups to retrieve the partition by key for the rebuilding process.

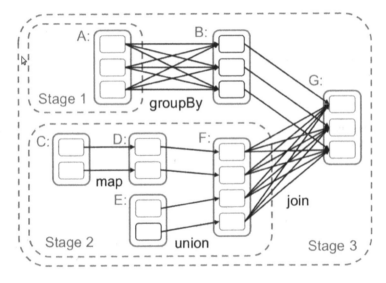

Fig. 3.4: Boxes RDDs. Black partitions are cached. While recovering RDD G stage 1, 2, and 3 need to be executed in that order. Since B is cached stage 1 is not needed. Only stage 2 and 3 are run [5].

3.3 Spark Installation

Spark provides fast and high-level APIs in Scala, Java, and Python that make writing parallel jobs easy. It also provides an optimized engine that supports graph computations. It also supports higher-level distributed tools like Shark, MLib, Bagel, and Spark Streaming [7].

3.3.1 Pre-requisites

Download Spark by visiting the project page in [7]. Java is the primary requirement to run Spark. Set the JAVA_HOME environment variable pointing to a Java installation.

This installation procedure assumes you have UNIX/LINUX-based operating system, i.e.,Ubuntu 14.04 is used in this tutorial.

Java Installation

```
user@spark:~$ java -version
java version "1.6.0_45"
```

```
Java(TM) SE Runtime Environment (build 1.6.0_45-b06)
Java HotSpot(TM) 64-Bit Server VM (build 20.45-b01, mixed mode)
```

If the above command fails then install java through any method, commonly installed via:

```
user@spark:~$ sudo apt-get install openjdk-6-jdk
```

Download and extract `Scala 2.10.0` from `scala-lang.org` and export the `bin` folder to the system `PATH` variable. Update the following to `~/.bashrc` file, replace `/path/to/` to the absolute path of the Scala downloaded earlier.

```
export SCALA_HOME="/path/to/scala-2.9.3/bin"
```

Spark uses `Simple Build Tool(SBT)` for compiling and managing source code. To compile the Spark source, in the top-level directory of the Spark source run the following command. The `assembly` function bundles all the packages and compiled source files in to an executable jar that is shared across the cluster.

```
user@spark:~/spark$ sbt/sbt assembly
[info] Loading project definition from /home/user/spark/project/
    project
[info] Loading project definition from /home/user/spark/project
[info] Set current project to root (in build file:/home/user/
    spark/)
[info] Updating {file:/home/user/spark/}core...
[info] Resolving org.apache.derby#derby;10.4.2.0 ...
[info] Done updating.
[info] Updating {file:/home/user/spark/}streaming...
.
.
.
Strategy 'concat' was applied to a file
[warn] Strategy 'discard' was applied to 3 files
[warn] Strategy 'filterDistinctLines' was applied to 2 files
[warn] Strategy 'first' was applied to 821 files
[info] Checking every *.class/*.jar file's SHA-1.
[info] Packaging /home/user/spark/examples/target/scala-2.9.3/
    spark-examples-assembly-0.8.1-incubating.jar ...
[info] Done packaging.
[success] Total time: 805 s, completed Jan 12, 2014 7:20:10 PM
```

Spark comes with several sample programs in the examples directory. To run one of the samples, use the `run-example` script in the top-level Spark directory.

```
./run-example <class> <params>
```

For example, try

```
./run-example org.apache.spark.examples.SparkPi local
```

Executing python-based examples use.

```
./spark-submit examples/src/main/python/pi.py 10
```

command line interface of scripts `run-example` and `spark-submit` takes the `class` and `<master>` parameter which indicates the cluster URL to launch Spark jobs. The value can be an URL where the cluster is running or `local` to run in the same system. `local[N]`, N denotes the number of threads launched locally.

Finally, you can run Spark interactively through modified versions of the Scala shell `./spark-shell` or Python interpreter `./pyspark`.

3.3.2 Getting Started

This section introduces Spark using a modified Scala shell, `spark-shell`. Since Spark is built using Scala languages the examples shown here are also in Scala (Note: Scala programming knowledge is not primary requirement for using Spark, you can pick it up using the shell).

Spark's shell provides powerful APIs to interactively analyze datasets. `spark-shell` launches a `scala>` prompt which pre-loads the Spark jar and other dependencies.

```
user@spark:~/spark$ ./spark-shell
14/09/02 12:13:45 INFO SecurityManager: Using Spark's default
    log4j profile: org/apache/spark/log4j-defaults.properties
14/09/02 12:13:45 INFO SecurityManager: Changing view acls to:
    mak
14/09/02 12:13:45 INFO SecurityManager: SecurityManager:
    authentication disabled; ui acls disabled; users with view
    permissions: Set(mak)
14/09/02 12:13:45 INFO HttpServer: Starting HTTP Server
Welcome to
      ____              __
     / __/__  ___ _____/ /__
    _\ \/ _ \/ _ `/ __/  '_/
   /___/ .__/\_,_/_/ /_/\_\   version 1.0.1
      /_/
```

```
Using Scala version 2.10.4 (OpenJDK 64-Bit Server VM,
    Java 1.7.0_55)
Type in expressions to have them evaluated.
Type :help for more information.
.
.
.
14/09/02 12:13:51 INFO HttpServer: Starting HTTP Server
14/09/02 12:13:52 INFO SparkUI: Started SparkUI at http
    ://192.168.1.4:4040
14/09/02 12:13:53 INFO Executor: Using REPL class URI: http
    ://192.168.1.4:36901
14/09/02 12:13:53 INFO SparkILoop: Created spark context..
Spark context available as sc.

scala>
```

Spark's data abstractions are Resilient Distributed Datasets (RDDs), which are a distributed collection of items partitioned across the cluster. These RDDs can be created and transformed, refer Section 3.2 for more details.

Create an RDD using *README* text file.

```
scala> val textFile = sc.textFile("README.MD")
textFile: org.apache.spark.rdd.RDD[String] = MappedRDD[1] at
    textFile at <console>:12
```

RDDs are bundled with actions that return values and transformations to new RDDs.

Counting the number of items in an textFile RDD.

```
scala> textFile.count()
res0: Long = 111
```

Returning the first item of textFile RDD.

```
scala> textFile.first()
res1: String = # Apache Spark
```

Applying filter transformation points to a new RDD, which is a subset of the original RDD.

Filter items in textFile RDD containing *Spark* and count the occurrences.

```
scala> val linesWithSpark = textFile.filter(line => line.
    contains("Spark"))
linesWithSpark: org.apache.spark.rdd.RDD[String] = FilteredRDD
    [2] at filter at <console>:14

scala> linesWithSpark.count()
res2: Long = 15
```

RDD transformations can be chained.

```
scala> val linesWithSpark = textFile.filter(line => line.
    contains("Spark")).count()
linesWithSpark: Long = 15
```

More RDD Operations

Several complex operations can be performed by chaining, to find the line with the most number of words.

```
scala> textFile.map(line => line.split(" ").size).reduce((a, b)
    => if (a > b) a else b)
res3: Int = 15
```

First each line of the dataset is split by ' ' (space) and counted to determine the number of words. This Line and count is passed in to a reduced closure that checks for greatest count in each line and returns the largest count, i.e. the line with the most number of words. All the language features of Scala/Java can be used as arguments to map and reduce functions. For example, we can use Scala built-in max() function.

```
scala> import java.lang.Math
import java.lang.Math

scala> textFile.map(line => line.split(" ").size).reduce((a, b)
    => Math.max(a, b))
res4: Int = 15
```

MapReduce can also be implemented using Spark.

```
scala> val wordCounts = textFile.flatMap(line => line.split(" ")).
    map(word => (word, 1)).reduceByKey((a, b) => a + b)
wordCounts: org.apache.spark.rdd.RDD[(java.lang.String, Int)] =
    MapPartitionsRDD[10] at reduceByKey at <console>:15

scala> wordCounts.collect()
res5: Array[(java.lang.String, Int)] = Array(("",121), (submitting
    ,1), (find,1), (versions,4), (making,1), (Regression,1), (via,2)
    , (required,1), (necessarily,1), (open,2), (packaged,1), (When
    ,1), (All,1), (code,,1), (requires,1), (SPARK_YARN=true,3),
    (guide](http://spark.incubator.apache.org/docs/latest/
    configuration.html),1), (take,1), (project,5), (no,1), (systems
    .,1), (YARN,1), (file,1), (<params>`.,1), (Or,,1), (`<
    dependencies>`,1), (About,1), (project's,3), (not,3), (programs
    ,2), (does,1), (given.,1), (artifact,1), (sbt/sbt,6), (`<master
    >`,1), (local[2],1), (runs.,1), (you,5), (building,1), (Along,1)
    , (Lightning-Fast,1), (incubation,2), ((ASF),,1), (Hadoop,,1),
    (use,1), (MRv2,,1), (it,2), (directory.,1), (overview,1), (The
    ,2), (easiest,1), (Note,1), (setup,1), ("org.apache.hadoop"...
```

This is a simple wordcount example. Here transformations like `flatMap`, `map`, and `reduceByKey` are used to compute counts per word in a file as an RDD of (String, Int) pairs. `collect` action is used to gather the RDD pairs from all the partitions.

Caching

Spark also provides in-memory storage of datasets which can used to perform operations repeatedly reducing the disk access overhead. Using the same `linesWith Spark` RDD discussed previously. `cache` action stores the contents of the RDD in-memory.

```
scala> linesWithSpark.cache()
res9: org.apache.spark.rdd.RDD[String] = FilteredRDD[11] at
    filter at <console>:15

scala> linesWithSpark.count()
res10: Long = 15
```

The advantage of Spark is realized when using the Spark shell to interactively analyze large datasets that are spread across hundreds of nodes.

Sample Applications

Spark is ideal in building several data-parallel applications. Here are some examples.

Console Log Mining: Consider a scenario where a system administrator needs to search terabytes of logs in the Hadoop Distributed File System (HDFS) to find the cause. Using Spark's RDD implementation the system admin can load only the error logs in to the memory for future operations. The operations is as follows. The filter operation only selects the lines in the dataset that begin with ERROR.

```
scala> lines = sc.textFile("hdfs://....")
scala> errors = lines.filter(_.startsWith("ERROR"))
scala> errors.cache()
```

An RDD is created from the file stored in HDFS as collection of lines. This RDD is further transformed in to another RDD containing only the lines that start with ERROR. To enable repeated actions, the RDD is cached in-memory.

This RDD can be subject to future transformations. The RDD with ERROR lines can be further filtered to check for errors of type HDFS. The resulting RDD can be used to collect time fields of these errors. The code for this is as follows.

```
scala> lines = sc.textFile("hdfs://....")
scala> errors = lines.filter(_.startsWith("ERROR"))
scala> errors.cache()
scala> errors.filter(_.contains("HDFS"))
            .map(_.split('\t')(3))
            .collect()
```

The process can be explained by Figure 3.5.

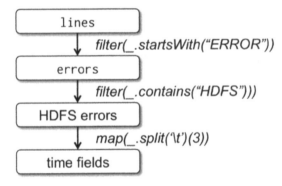

Fig. 3.5: The rectangles represent transformed RDDs, arrows represent transformations [7].

MapReduce using RDDs: Spark RDDs can be used in MapReduce operations. Given two functions, that resembles the map and reduce functional programming principles. Given elements of the dataset of type T.

- myMap : $T \Rightarrow List[(K_i , V_i)]$

- myReduce : $(K_i, List[V_i]) \Rightarrow List[R]$

These two operations can be combined as.

```
scalal> data.flatMap(myMap)
    .groupByKey()
    .map((k, vs) => myReduce(k, vs))
```

If a user-defined combiner function, myCombiner, is also available, it can be plugged in place of the generic groupByKey API.

```
scala> data.flatMap(myMap)
    .reduceByKey(myCombiner)
    .map((k, v) => myReduce(k, v))
```

Note: This example assumes you have a data variable defined.

3.3.3 Example: Scala Application

Scala example

Lets call the simple Spark application SimpleApp.scala.

Step 1: Create a directory structure for the application `src` and `target` directories.

```
user@spark:~/spark$ mkdir simple
user@spark:~/spark/simple$ mkdir -p src/main/scala
```

Step 2: This file depends on the Spark API, and also includes an sbt configuration file, `simple.sbt` which explains that Spark is a dependency. This file also adds a repository that Spark depends on:

```
user@spark:~/spark/simple$ cat simple.sbt
name := "Simple Project"

version := "1.0"

scalaVersion := "2.10.1"

libraryDependencies += "org.apache.spark" %% "spark-core" %
    "1.0.1"

resolvers += "Akka Repository" at "http://repo.akka.io/releases/"
```

Step 3: Write a spark program to count the number of a's and b's in a user-defined text file. Create a file `SimpleApp.scala` in `src/main/scala` directory with the following code:

```scala
/*** SimpleApp.scala ***/
import org.apache.spark.SparkContext
import org.apache.spark.SparkContext._

object SimpleApp {
  def main(args: Array[String]) {
    val logFile = "README.md"
    val conf = new SparkConf().setAppName("SimpleApp")
    val sc = new SparkContext(conf)
    val logData = sc.textFile(logFile, 2).cache()
    val numAs = logData.filter(line =>
                        line.contains("a")).count()
    val numBs = logData.filter(line =>
                        line.contains("b")).count()
    println("Lines with a: %s,
            Lines with b: %s".format(numAs, numBs))
  }
```

In this example, unlike other examples discussed before with the Spark shell which initializes its own `SparkContext`, it is initialized programatically. To initialize the `SparkContext`, an instance of the `SparkConf` is created with application name set using `setAppName` method [7]. An RDD `logData` is created using the text file `logFile`. To count the number of a's and b's `filter` action is applied followed by a `count` action on the RDD.

Step 4: Check if the organization of the project is as below

```
user@spark:~/spark/simple$ find .
.
./simple.sbt
./src
./src/main
./src/main/scala
./src/main/scala/SimpleApp.scala
```

Step 5: The application is packaged in to a jar file that is deployed across the clusters and run.

```
user@spark:~/spark/simple$ sbt package
[info] Set current project to Simple Project (in build file:/app
    /spark/spark-1.0.1/simple/)
[info] Updating {file:/app/spark/spark-1.0.1/simple/}simple...
[info] Resolving org.eclipse.jetty.orbit#javax.transaction
    ;1.1.1.v201105210645 .[info] Resolving org.eclipse.jetty.
    orbit#javax.mail.glassfish;1.4.1.v20100508202[info]
    Resolving org.eclipse.jetty.orbit#javax.activation;1.1.0.
    v201105071233 ..[info] Resolving org.spark-project.akka#akka
    -remote_2.10;2.2.3-shaded-protobuf .[info] Resolving org.
    spark-project.akka#akka-actor_2.10;2.2.3-shaded-protobuf ..[
    info] Resolving org.spark-project.akka#akka-slf4j_2
    .10;2.2.3-shaded-protobuf ..[info] Resolving org.fusesource.
    jansi#jansi;1.4 ...
[info] Done updating.
[info] Compiling 1 Scala source to /app/spark/spark-1.0.1/simple
    /target/scala-2.10/classes...
[info] Packaging /app/spark/spark-1.0.1/simple/target/scala
    -2.10/simple-project_2.10-1.0.jar ...
[info] Done packaging.
[success] Total time: 14 s, completed 2 Sep, 2014 4:16:16 PM
```

Step 6: Submit the application to the Spark cluster using the `bin/spark-submit` script. It takes 3 arguments

- *class*: The name of the class that needs to be executed

- *master*: Cluster URL or local execution, local[N] denoting number of threads to be used

- Project jar file contains Spark jar and project class files bundled. This jar is shared with all the nodes in the cluster.

```
user@spark:~/spark/simple$ ../bin/spark-submit --class "
    SimpleApp" --master local target/scala-2.10/simple-project_2
    .10-1.0.jar
14/09/02 16:17:16 INFO Utils: Using Spark's default log4j
    profile: org/apache/spark/log4j-defaults.properties
```

```
14/09/02 16:17:16 WARN Utils: Your hostname, thinkpad resolves
    to a loopback address: 127.0.0.1; using 192.168.1.4 instead
    (on interface wlan0)
14/09/02 16:17:16 WARN Utils: Set SPARK_LOCAL_IP if you need to
    bind to another address
14/09/02 16:17:16 INFO SecurityManager: Changing view acls to:
    mak
14/09/02 16:17:16 INFO SecurityManager: SecurityManager:
    authentication disabled; ui acls disabled; users with view
    permissions: Set(mak)
14/09/02 16:17:17 INFO Slf4jLogger: Slf4jLogger started
14/09/02 16:17:17 INFO Remoting: Starting remoting
.
.
.
14/09/02 16:17:20 INFO TaskSetManager: Finished TID 3 in 8 ms on
    localhost (progress: 2/2)
14/09/02 16:17:20 INFO TaskSchedulerImpl: Removed TaskSet 1.0,
    whose tasks have all completed, from pool
14/09/02 16:17:20 INFO DAGScheduler: Stage 1 (count at simpleapp
    .scala:12) finished in 0.022 s
14/09/02 16:17:20 INFO SparkContext: Job finished: count at
    simpleapp.scala:12, took 0.029117467 s
Lines with a: 73, Lines with b: 35
```

3.3.4 Spark with Python

There are many differences between Spark Python and Spark Scala APIs, some of the key differences are:

Python is a dynamically typed language and RDDs have the ability to store objects of multiple types. They support the same methods as the Scala APIs but take Python functions and return collections. Python uses anonymous functions called lambda functions that can be passed as arguments to the API.

For example, a function to check "ERROR" present in a line can be written as:

```
def is_error(line):
    return 'ERROR' in line
errors = logData.filter(is_error)
```

the lambda alternative is:

```
logData = sc.textFile(logFile).cache()
errors = logData.filter(lambda line: 'ERROR' in line)
```

Class instances will be serialized similar to Scala objects and these lambda functions are passed to workers along with any reference objects.

Similar to the Spark interactive shell, PySpark also provides an easy to use interactive shell.

`PySpark` depends on version 2.6 or higher, they are executed using a CPython interpreter that accommodates for the C extensions. PySpark uses the installed python version on the system by default, any new version can be updated for use by updating the value of `PYSPARK_PYTHON` environmental variable in `conf/spark-env.sh`

Running the `pyspark` script, launches Python interpreter that is changed to run PySpark applications. Before using PySpark, Spark must be assembled. Check section 3.3.1 for instructions.

```
$ ./pyspark
14/09/04 21:33:36 INFO HttpServer: Starting HTTP Server
14/09/04 21:33:36 INFO SparkUI: Started SparkUI at http
    ://192.168.1.4:4040
Welcome to
      ____              __
     / __/__  ___ _____/ /__
    _\ \/ _ \/ _ `/ __/  '_/
   /__ / .__/\_,_/_/ /_/\_\   version 1.0.1
      /_/

Using Python version 2.7.6 (default, Mar 22 2014 22:59:56)
SparkContext available as sc.
>>>
```

The PySpark shell can be used right away to interact with data with simple to use API:

For example, the procedure to read a file and filter out the lines that contain the word "spar" and select the first 5 lines can be implemented as follows.

```
>>> words = sc.textFile("/usr/share/dict/words")
>>> words.filter(lambda w: w.startswith("spar")).take(5)
[u'spar', u'sparable', u'sparada', u'sparadrap', u'sparagrass']
>>> help(pyspark) # Show all pyspark functions
```

Similar to the Spark Scala shell, the pyspark shell creates a SparkContext that provides all the API to run applications locally by default. To connect to a cluster, similar to Scala-shell, remotely or standalone:

```
$ MASTER=spark://IP:PORT ./pyspark
```

Or, to use four cores on the local machine:

```
$ MASTER=local[4] ./pyspark
```

3.3.5 Example: Python Application

Step 1: Counting the number lines containing a's and b's using Spark python APIs. An instance of `SparkConf` with the name of the applications set with `setAppName` attributed. This `conf` is then passed as an argument to `Spark Context`. Using the `textFile` API an RDD of the file *README.md* is created. A python lambda function is created as below to count the number of lines that contains a's and b's.

```python
from pyspark import SparkContext, SparkConf

logFile = "README.md"
conf = SparkConf().setAppName("Pythonlines")
sc = SparkContext(conf=conf)
logData = sc.textFile(logFile).cache()

numAs = logData.filter(lambda s: 'a' in s).count()
numBs = logData.filter(lambda s: 'b' in s).count()

print "Lines with a: " + str(numAs) +
              ", lines with b: " + str(numBs)
```

Step 2: Run this example using `bin/spark-submit` script

```
user@spark:~/spark/simple$ ../bin/spark-submit lines.py
14/09/04 22:30:04 INFO Utils: Using Spark's default log4j
    profile: org/apache/spark/log4j-defaults.properties
14/09/04 22:30:04 WARN Utils: Your hostname, thinkpad resolves
    to a loopback address: 127.0.0.1; using 192.168.1.4 instead
    (on interface wlan0)
14/09/04 22:30:04 WARN Utils: Set SPARK_LOCAL_IP if you need to
    bind to another address
14/09/04 22:30:04 INFO SecurityManager: Changing view acls to:
    mak
14/09/04 22:30:04 INFO SecurityManager: SecurityManager:
    authentication disabled; ui acls disabled; users with view
    permissions: Set(mak)
14/09/04 22:30:05 INFO Slf4jLogger: Slf4jLogger started
14/09/04 22:30:05 INFO Remoting: Starting remoting
14/09/04 22:30:05 INFO Remoting: Remoting started; listening on
    addresses :[akka.tcp://spark@192.168.1.4:43599]
14/09/04 22:30:05 INFO Remoting: Remoting now listens on
    addresses: [akka.tcp://spark@192.168.1.4:43599]
...
14/09/04 22:30:07 INFO TaskSchedulerImpl: Removed TaskSet 1.0,
    whose tasks have all completed, from pool
14/09/04 22:30:07 INFO DAGScheduler: Stage 1 (count at /app/
    spark/spark-1.0.1/simple/lines.py:8) finished in 0.022 s
14/09/04 22:30:07 INFO SparkContext: Job finished: count at /app
    /spark/spark-1.0.1/simple/lines.py:8, took 0.033365716 s
Lines with a: 73, lines with b: 35
```

3.4 Deploying Spark

The Spark driver program uses the `SparkContext` object to coordinate between independent processes on a cluster. The `SparkContext` can connect to a variety of cluster managers (Standalone or Mesos/YARN) which allocate resources to applications. Once a connection is established between the driver program and the nodes, worker processes are launched to run computations and store data. The application code bundled into a JAR or python files are passed to SparkContext. Finally, the SparkContext sends the tasks to the workers to run (Figure 3.6).

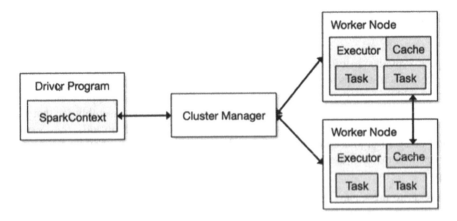

Fig. 3.6: Application execution flow [7]

Several advantages to using this architectures.

- Each application receives its own set of worker processes which are available for the duration of the applications and also run tasks in multiple threads. This allows for isolating application on both scheduling and execution aspects. However, these applications cannot share data between them without writing to an external storage system.

- Spark is also agnostic to the underlying cluster management system, it is relatively easy to run Spark even on the managers with other clusters as long as it can recognize the worked processes.

- The driver program schedules the tasks on the cluster such that it runs very close to the data, generally on the same local area network. Its better to open a remote procedure call to driver to submit operations near the data than far away from the workers.

The Spark system currently supports three cluster managers:

- **Standalone:** a simple cluster manager included with Spark that makes it easy to set up a cluster.

- **Apache Mesos:** a general cluster manager that can also run Hadoop `MapReduce` and service applications.

- **Hadoop YARN:** Yet another resource manager for Hadoop version 2.

In addition, Sparks EC2 launch scripts make it easy to launch a standalone cluster on **Amazon EC2** .

3.4.1 Submitting Applications

Spark provides an efficient method to submit applications using `bin/spark-submit` script. This script is enough to manage all Spark supported cluster managers without any extra configuration changes. For Scala-based applications, any dependencies are assembled alongside the application as an assembly jar that is copied to the workers. For Python users `--py-files` are used to add `.py`, `.egg` and `.zip` files. These files are copied to the working directories of all the workers. Over time this can use a significant amount of space and will need to be cleaned up.

This script takes care of classpath setting of Spark and its dependencies and supports various deploy modes, `bin/spark-submit --help` gives the following:

```
./bin/spark-submit \
  --class <main-class>
  --master <master-url> \
  --deploy-mode <deploy-mode> \
  ... # other options
  <application-jar> \
  [application-arguments]
```

Commonly used options are:

- `--class`: This argument indicates the entry point of the applications (e.g. org.apache.spark.examples.SparkPi)

- `--master`: URL of the cluster master node.

- `--deploy-mode`: Deploy locally or on the cluster

- `application-jar`: Globally visible path to the jar containing the application and its dependencies.

Different deployment use-cases using `spark-shell` script:

```
# Run application locally on 8 cores
./bin/spark-submit \
  --class org.apache.spark.examples.SparkPi \
  --master local[8] \
  /path/to/examples.jar \
  100

# Run on a Spark standalone cluster
./bin/spark-submit \
  --class org.apache.spark.examples.SparkPi \
  --master spark://192.168.1.100:7077 \
  --executor-memory 20G \
  --total-executor-cores 100 \
  /path/to/examples.jar \
  1000

# Run on a YARN cluster
export HADOOP_CONF_DIR=XXX
./bin/spark-submit \
  --class org.apache.spark.examples.SparkPi \
  --master yarn-cluster \
  --executor-memory 20G \
  --num-executors 50 \
  /path/to/examples.jar \
  1000

# Run a Python application on a cluster
./bin/spark-submit \
  --master spark://192.168.1.100:7077 \
  examples/src/main/python/pi.py \
  1000
```

3.4.2 Standalone Mode

Starting a Cluster Manually

You can start a standalone master server by executing:

```
user@spark:~/spark$ ./sbin/start-master.sh
starting org.apache.spark.deploy.master.Master, logging to /home
    /user/spark/bin/../logs/spark-user-org.apache.spark.deploy.
    master.Master-1-spark.out
```

Spark URL of the master can be seen on http://localhost:8080 which is
the web-ui of the cluster (Figure 3.7). For example, in this case
spark://HackStation:7077. (HackStation is the hostname)

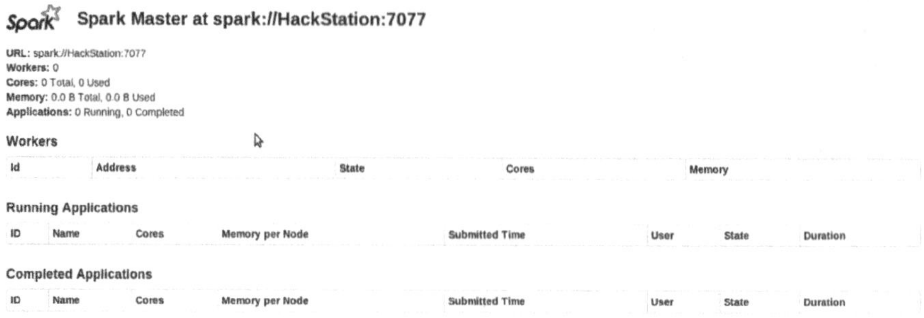

Spark Spark Master at spark://HackStation:7077

URL: spark://HackStation:7077
Workers: 0
Cores: 0 Total, 0 Used
Memory: 0.0 B Total, 0.0 B Used
Applications: 0 Running, 0 Completed

Workers

Id	Address	State	Cores	Memory

Running Applications

ID	Name	Cores	Memory per Node	Submitted Time	User	State	Duration

Completed Applications

ID	Name	Cores	Memory per Node	Submitted Time	User	State	Duration

Fig. 3.7: Spark Master Web UI

One or more workers can be connected to the master executing the following:

```
user@spark:~/spark$ ./spark-class org.apache.spark.deploy.worker
    .Worker spark://HackStation:7077

SLF4J: Class path contains multiple SLF4J bindings.
SLF4J: Found binding in [jar:file:/app/spark/spark-0.8.1/tools/
    target/scala-2.9.3/spark-tools-assembly-0.8.1-incubating.jar
    !/org/slf4j/impl/StaticLoggerBinder.class]
SLF4J: Found binding in [jar:file:/app/spark/spark-0.8.1/
    assembly/target/scala-2.9.3/spark-assembly-0.8.1-incubating-
    hadoop1.0.4.jar!/org/slf4j/impl/StaticLoggerBinder.class]
.
.
.
```

Once you have started a worker, look at the masters web UI (http://localhost:
8080 by default). You should see the new node listed there, along with its number
of cores and memory (Figure 3.8).

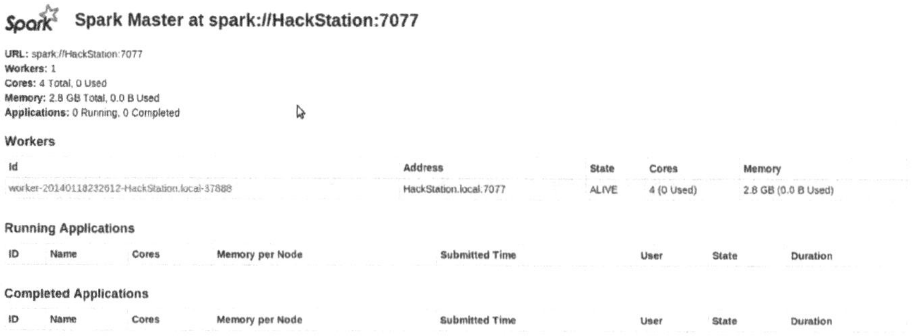

Spark Spark Master at spark://HackStation:7077

URL: spark://HackStation:7077
Workers: 1
Cores: 4 Total, 0 Used
Memory: 2.8 GB Total, 0.0 B Used
Applications: 0 Running, 0 Completed

Workers

Id	Address	State	Cores	Memory
worker-20140118232612-HackStation.local-57888	HackStation.local.7077	ALIVE	4 (0 Used)	2.8 GB (0.0 B Used)

Running Applications

ID	Name	Cores	Memory per Node	Submitted Time	User	State	Duration

Completed Applications

ID	Name	Cores	Memory per Node	Submitted Time	User	State	Duration

Fig. 3.8: Spark Worker Web UI

Finally, the following configuration options can be passed to the master and worker as shown in Table 3.2.

Table 3.2: Configuration options

Argument	Meaning
`-i IP, --ip IP`	IP address or DNS name to listen on
`-p PORT, --port PORT`	Port for service to listen on (default: 7077 for master, random for worker)
`--webui-port PORT`	Port for web UI (default: 8080 for master, 8081 for worker)
`-c CORES, --cores CORES`	Total CPU cores allowed for Spark applications. By default all are available on the workers.
`-m MEM, --memory MEM`	Total amount of memory allowed for Spark applications, by default the system's memory minus 1 GB)
`-d DIR, --work-dir DIR`	Directory used to store job and output logs, default: `SPARK_HOME/work`

Cluster Launch Scripts

To use the launch scripts for a Spark standalone cluster, a file `conf/slaves` is created. This file contains all the hostnames of the worker nodes. The master machine is given access to the worker nodes by way of creating a password-less sh connection. For more details on this refer Pre-requisites section in Chapter 2. For testing purposes `localhost` can be added to the list then the master and worker will be the same node (Single node cluster).

The scripts are located at the `SPARK_HOME/bin`. These scripts function very similar to the Hadoop launch scripts discussed in Chapter 2.

- `bin/start-master.sh` - Starts a master instance on the machine the script is executed on.

- `bin/start-slaves.sh` - Starts a slave/worker instance on each machine specified in the *conf/slaves* file.

- `bin/start-all.sh` - Starts both a master and a number of slaves as described above.

- `bin/stop-master.sh` - Stops the master that was started via the *bin/start-master.sh* script.

- `bin/stop-slaves.sh` - Stops the slave instances that were started via *bin/start-slaves.sh*.

- `bin/stop-all.sh` - Stops both the master and the slaves as described above.

Note: These scripts must be executed on the machine in which you want to run the Spark master on, but not in your local machine.

Submitting an Application to the Cluster

Using the code developed in section 3.3.3. The code can be submitted to the cluster launched at `spark://HackStation:7077`.

```
../bin/spark-submit --class "SimpleApp" --master spark://
    HackStation:7077 target/scala-2.10/simple-project_2.10-1.0.
    jar
14/09/02 19:19:40 INFO Utils: Using Spark's default log4j
    profile: org/apache/spark/log4j-defaults.properties
14/09/02 19:19:40 WARN Utils: Your hostname, thinkpad resolves
    to a loopback address: 127.0.0.1; using 192.168.1.4 instead
    (on interface wlan0)
14/09/02 19:19:40 WARN Utils: Set SPARK_LOCAL_IP if you need to
    bind to another address
.
.
.
14/09/02 19:19:57 INFO DAGScheduler: Completed ResultTask(1, 1)
14/09/02 19:19:57 INFO TaskSetManager: Finished TID 3 in 28 ms
    on 192.168.1.4 (progress: 2/2)
14/09/02 19:19:57 INFO DAGScheduler: Stage 1 (count at simpleapp
    .scala:12) finished in 0.030 s
14/09/02 19:19:57 INFO TaskSchedulerImpl: Removed TaskSet 1.0,
    whose tasks have all completed, from pool
14/09/02 19:19:57 INFO SparkContext: Job finished: count at
    simpleapp.scala:12, took 0.046617489 s
Lines with a: 73, Lines with b: 35
```

Check the `http://localhost:8080/` to see progress of the Job (Figure 3.9).

Fig. 3.9: Spark Application Progress

References

1. J. Dean and S. Ghemawat. MapReduce: Simplified data processing on large clusters. Commun. ACM, 51(1):107-113, 2008.
2. H. Yang, A. Dasdan, R. Hsiao, and D. S. Parker. Map-reduce-merge: simplified relational data processing on large clusters. In SIGMOD 07, pages 1029-1040. ACM, 2007.
3. M. Isard, M. Budiu, Y. Yu, A. Birrell, and D. Fetterly. Dryad: Distributed data-parallel programs from sequential building blocks. In EuroSys 2007, pages 59-72, 2007.
4. B. Hindman, A. Konwinski, M. Zaharia, and I. Stoica. A common substrate for cluster computing. In Workshop on Hot Topics in Cloud Computing (HotCloud) 2009, 2009.
5. Zaharia, Matei, et al. "Spark: cluster computing with working sets." Proceedings of the 2nd USENIX conference on Hot topics in cloud computing. 2010.
6. C. Olston, B. Reed, U. Srivastava, R. Kumar, and A. Tomkins. Pig latin: a not-so-foreign language for data processing. In *SIGMOD 08. ACM*, 2008.
7. Apache Hive. http://hadoop.apache.org/hive
8. Li, Kai. "IVY: A Shared Virtual Memory System for Parallel Computing." ICPP (2). 1988.
9. B. Nitzberg and V. Lo. Distributed shared memory: a survey of issues and algorithms. Computer, 24(8):52-60, aug 1991.
10. A.-M. Kermarrec, G. Cabillic, A. Gefflaut, C. Morin, and I. Puaut. A recoverable distributed shared memory integrating coherence and recoverability. In FTCS 95. IEEE Computer Society, 1995.
11. R. Bose and J. Frew. Lineage retrieval for scientific data processing: a survey. ACM Computing Surveys, 37:128, 2005.
12. J. B. Carter, J. K. Bennett, and W. Zwaenepoel. Implementation and performance of Munin. In SOSP 91. ACM, 1991.
13. D. Gelernter. Generative communication in linda. ACM Trans. Program. Lang. Syst., 7(1):80-112, 1985.
14. B. Liskov, A. Adya, M. Castro, S. Ghemawat, R. Gruber, U. Maheshwari, A. C. Myers, M. Day, and L. Shrira. Safe and efficient sharing of persistent objects in thor. In SIGMOD 96, pages 318-329. ACM, 1996.
15. G. Malewicz, M. H. Austern, A. J. Bik, J. C. Dehnert, I. Horn, N. Leiser, and G. Czajkowski. Pregel: a system for large-scale graph processing. In SIGMOD, pages 135146, 2010.
16. J. Ekanayake, H. Li, B. Zhang, T. Gunarathne, S.-H. Bae, J. Qiu, and G. Fox. Twister: a runtime for iterative mapreduce. In HPDC 10, 2010.
17. Y. Bu, B. Howe, M. Balazinska, and M. D. Ernst. HaLoop: efficient iterative data processing on large clusters. Proc. VLDB Endow., 3:285-296, September 2010.
18. Scala programming language. http://www.scala-lang.org.
19. B. Hindman, A. Konwinski, M. Zaharia, A. Ghodsi, A. D. Joseph, R. H. Katz, S. Shenker, and I. Stoica. Mesos: A platform for fine-grained resource sharing in the data center. Technical Report UCB/EECS-2010-87, EECS Department, University of California, Berkeley, May 2010.
20. M. Zaharia, D. Borthakur, J. Sen Sarma, K. Elmeleegy, S. Shenker, and I. Stoica. Delay scheduling: A simple technique for achieving locality and fairness in cluster scheduling. In EuroSys 2010, April 2010.
21. R. Power and J. Li. Piccolo: Building fast, distributed programs with partitioned tables. In Proc. OSDI 2010, 2010.
22. Spark, Apache. [Online] Available: http://spark.incubator.apache.org/docs/latest/

Chapter 4

Programming Internals of Scalding and Spark

4.1 Scalding

Scalding is a Scala-based library built on top of *Cascading* , a Java library that forms an abstraction over low-level Hadoop API. It is comparable to Pig, but brings the advantages of Scala in building MapReduce jobs [1].

4.1.1 Installation

The installation procedure is as simple as downloading the source code from the official Twitter Github repository and compiling the project to build the scalding jar. Scalding version 0.10.0 is used throughout this book.

Step 1: Download the project zip file from

```
https://github.com/twitter/scalding
```

Step 2: Scalding project comes with the simple build tool (sbt) that has all the configuration details for building and testing the scalding project.

Update the project to receive any new changes as it is always under development and also the dependencies.

```
user@ubuntu:~/scalding$ sbt update
[info] Loading project definition from /app/scalding/scalding
    -0.10.0/project
[info] Set current project to scalding (in build file:/app/
    scalding/scalding-0.10.0/)
[info] Updating {file:/app/scalding/scalding-0.10.0/}scalding-
    date...
```

© Springer International Publishing Switzerland 2015
K.G. Srinivasa and A.K. Muppalla, *Guide to High Performance
Distributed Computing*, Computer Communications and Networks,
DOI 10.1007/978-3-319-13497-0_4

```
[info] Updating {file:/app/scalding/scalding-0.10.0/}scalding-
    args...
[info] Updating {file:/app/scalding/scalding-0.10.0/}maple...
[info] Resolving org.fusesource.jansi#jansi;1.4 ...
[info] Done updating.
[info] Resolving org.apache.mahout#mahout;0.9 ...

....

[info] Updating {file:/app/scalding/scalding-0.10.0/}scalding-
    repl...
[info] Resolving org.fusesource.jansi#jansi;1.4 ...
[info] Done updating.
[info] Resolving com.fasterxml.jackson.module#jackson-module-
    scala_2.9.3;2.2.3 .[info] Resolving org.fusesource.jansi#
    jansi;1.4 ...
[info] Done updating.
[info] Resolving org.codehaus.jackson#jackson-core-asl;1.8.8 ...
[info] Done updating.
[info] Resolving org.fusesource.jansi#jansi;1.4 ...
[info] Done updating.
[success] Total time: 11 s, completed 3 Sep, 2014 3:24:09 AM
```

Step 3: Run tests

```
user@ubuntu:~/scalding$ sbt test
[info] Compiling 2 Scala sources to /app/scalding/scalding-new/
    scalding-args/target/scala-2.9.3/test-classes...
[info] Compiling 3 Scala sources to /app/scalding/scalding-new/
    scalding-date/target/scala-2.9.3/test-classes...
[info] Compiling 85 Scala sources to /app/scalding/scalding-new/
    scalding-core/target/scala-2.9.3/classes...
[info] + Tool.parseArgs should
[info]    + handle the empty list
[info]    + accept any number of dashed args
[info]    + remove empty args in lists
[info]    + put initial args into the empty key
[info]    + allow any number of args per key
[info]    + allow any number of dashes
[info]    + round trip to/from string
[info]    + handle positional arguments
[info]    + handle negative numbers in args
[info]    + handle strange characters in the args
[info]    + access positional arguments using apply
[info]    + verify that args belong to an accepted key set
....
[info] Passed: Total 21, Failed 0, Errors 0, Passed 21
....
```

Step 4: Create a jar file with all the dependencies and the project executables. This is used when launching the `scald.rb` script for running Scalding jobs.

Execute the following to setup the required jar files:

```
user@ubuntu:~/scalding$ sbt assembly
[info] Loading project definition from /app/scalding/scalding-
    new/project
[info] Set current project to scalding (in build file:/app/
    scalding/scalding-new/)
[info] Including from cache: cascading-hadoop-2.5.2.jar
[info] Including from cache: cascading-core-2.5.2.jar
[info] Including from cache: riffle-0.1-dev.jar
[info] Including from cache: jgrapht-jdk1.6-0.8.1.jar
[info] Including from cache: janino-2.6.1.jar
[info] Including from cache: commons-compiler-2.6.1.jar
[info] Including from cache: scala-library-2.9.3.jar
....
[info] Assembly up to date: /app/scalding/scalding-new/target/
    scala-2.9.3/scalding-assembly-0.9.1.jar
[info] Assembly up to date: /app/scalding/scalding-new/scalding-
    args/target/scala-2.9.3/scalding-args-assembly-0.9.1.jar
....
```

Step 5: Write a simple Scalding wordcount program and test the functionality.

```
import com.twitter.scalding._

class WordCountJob(args : Args) extends Job(args) {
  TextLine( args("input") )
    .flatMap('line -> 'word){ line : String =>
                      line.split("""\s+""") }
    .groupBy('word) { _.size }
    .write( Tsv( args("output") ) )
}
```

The API level details will be explained in the latter sections of this chapter.

To run the job, copy the source code above into a `WordCountJob.scala` file, create a file named `someInputfile.txt` containing some arbitrary text, and then enter the following command from the root of the Scalding repository:

```
user@ubuntu:~/scalding$ scripts/scald.rb --local WordCountJob.
    scala --input someInputfile.txt --output ./someOutputFile.
    tsv
downloading hadoop-core-1.0.3.jar from http://repo1.maven.org/
    maven2/org/apache/hadoop/hadoop-core/1.0.3/hadoop-core
    -1.0.3.jar...
Successfully downloaded hadoop-core-1.0.3.jar!
downloading slf4j-log4j12-1.6.6.jar from http://repo1.maven.org/
    maven2/org/slf4j/slf4j-log4j12/1.6.6/slf4j-log4j12-1.6.6.jar
    ...
```

```
.
.
.
11:19:45 INFO flow.Flow: [WordCountJob] starting
11:19:45 INFO flow.Flow: [WordCountJob]  source: FileTap["
    TextLine[['offset', 'line']->[ALL]]"]["in"]
11:19:45 INFO flow.Flow: [WordCountJob]  sink: FileTap["
    TextDelimited[[UNKNOWN]->[ALL]]"]["haha"]
11:19:45 INFO flow.Flow: [WordCountJob]  parallel execution is
    enabled: true
11:19:45 INFO flow.Flow: [WordCountJob]  starting jobs: 1
11:19:45 INFO flow.Flow: [WordCountJob]  allocating threads: 1
11:19:45 INFO flow.FlowStep: [WordCountJob] starting step: local
11:19:45 INFO assembly.AggregateBy: using threshold value:
    100000
```

This runs the WordCount job in local mode (i.e., not on a Hadoop cluster). After a few seconds, your first Scalding job should be done.

```
user@ubuntu:~/scalding$ cat someOutputFile.tsv
Appendix    1
BASIS,  1
CONDITIONS   3
Contribution     3
Contribution(s) 3
Contribution." 1
Contributions)  1
Contributions.  2
Contributor 8
.
.
.
```

Note: Scalding works with Scala 2.9.3 and 2.10.0+ though a few configuration files must be changed for this to work. The file project/Build.scala has all the configuration settings for the Scalding project.

4.1.2 Programming Guide

Scalding is Scala wrapper built on top of Cascading library [3]. The Cascading library has many APIs for data processing, process scheduling, etc. Cascading abstracts the underlying Hadoop Map and Reduce tasks and also leverages the scalability of Hadoop. Developers bundle their programs as part of Cascading jar file that is distributed across the Hadoop cluster. Programs are written adhering to the **source–pipe–sink** paradigm. Data is captured from *sources*, passed through reusable *pipes* that define complex data analytic processes, and results are passed to *sinks* that is usually external storage. This is termed as a Cascading *flow*. Flows are grouped together to form "cascades" and these flows are executed provided all the dependencies are met. *Pipes* and *Flows* can be reused for different business needs [4].

Developers can write Cascading programs using Java and these are also viable for integration with larger applications.

Scalding provides large number of API that aid in writing efficient MapReduce programs. Some of the APIs along with examples are discussed below.

Note: For all the examples below the dataset given is stored in a file with name `in` and fed into the program using the command shown in the `run` step.

1. **read and write** The basic read and write functionality is responsible to reading from the source and writing to a sink. This is mandatory to establish a flow and create and process pipes.

```
import com.twitter.scalding._
import com.twitter.scalding.mathematics._
class testJob(args: Args) extends Job(args) {
    val input = TextLine("README.md")
        .read
        .write(TextLine("readwrite.txt"))
}
```

2. **map:** The general API usage is

```
pipe.map(currentFields -> newFields){function}
```

Pipe represents an input from source, `function` is applied on each element of the pipe.

Data: Consider sample dataset of currencies in USD, which need to be converted to INR (Table 4.1).

Table 4.1: Currency in USD

Currency in $
50
10
20
40

Code: Apply the `map` function on each element and multiply it by the exchange value of 60.

```
import com.twitter.scalding._
class testJob(args: Args) extends Job(args) {
```

```
    val input = Tsv("input")
        .map
        .write(Tsv("output"))
}
```

Run: Execute the job using `scald.rb` script.

```
$ ./scald.rb --local map.scala
```

Output: Output looks like as in Table 4.2

Table 4.2: Converted currency in INR

Currency in \$	Currency in INR
50	3000
10	600
20	1200
40	2400

3. **flatMap:** The general API usage is

```
pipe.flatMap(currentFields -> newFields){function}
```

Data: Consider a text file with line

```
I love scalding
```

Code: `flatMap` maps each element to a list and then flattens the list, emitting a tuple per item in the returned list

```
import com.twitter.scalding._

class testJob(args: Args) extends Job(args) {
    val input = Tsv(args("input"), 'lines)
        .flatMap('lines -> 'words){
            lines : String => lines.split(' ')
        }
        .write(Tsv("output.tsv"))
}
```

Run: using `scald.rb`

```
$ ./scald.rb --local flatMap.scala --input in
```

Output: Output looks like

```
I love Scalding I
I love Scalding love
I love Scalding Scalding
```

4. **mapTo :** It is equivalent to performing a map operation and then projecting only the newFields. The API usage is

```
pipe.mapTo(currentFields -> newFields){function}
```

Data: Consider a dataset of commodity prices and discounts, calculate the savings for each commodity. First column is price and second column is discount.

```
12000 200
350    25
4500   150
60     6
```

Code: Savings are determined by calculating price - discount for each row.

```
import com.twitter.scalding._

class testJob(args: Args) extends Job(args) {
    val input = Tsv(args("input"), ('price,'discount))
        .mapTo(('price, 'discount) -> ('savings)){
            x : (Int, Int) =>
            val(price, discount) = x
            (price - discount)
        }
        .write(Tsv("output.tsv"))
}
```

Alternative use of **map** and **project**

```
import com.twitter.scalding._

class testJob(args: Args) extends Job(args) {
    val input = Tsv(args("input"), ('price,'discount))
        .map(('price, 'discount) -> ('savings)){
            x : (Int, Int) =>
            val(price, discount) = x
            (price - discount)
        }
        .project('savings)
        .write(Tsv("output.tsv"))
}
```

Run: using scald.rb

```
../scald.rb --local mapTo.scala --input in
```

Output: Output in *output.tsv*

```
11800
325
4350
54
```

5. **flatMapTo**: The API usage is

```
pipe.flatMapTo(currentFields -> newFields){function}
```

It is similar to combining **flatMap** and **mapTo**

Data: Consider a string

```
I love Scalding
```

Code:

```
import com.twitter.scalding._

class testJob(args: Args) extends Job(args) {
    val input = Tsv(args("input"), 'lines)
        .flatMapTo('lines -> 'word){
            line : String => line.split(' ')
        }
        .write(Tsv("output.tsv"))
}
```

Run: using scald.rb

```
$ ./scald.rb --local flatMapTo.scala --input in
```

Output: Output looks like

```
I
love
Scalding
```

6. **project**: The API usage is

```
pipe.project(fields)
```

The output consists only of the fields listing as arguments.

Data: Consider a dataset of people and their ages. First column is the person name and second column is the age.

```
Anil        23
Kumar       45
Srinivasa   37
Brad        50
```

Code:

```
import com.twitter.scalding._

class testJob(args: Args) extends Job(args) {
    val input = Tsv(args("input"), ('name, 'age))
        .project('name)
        .write(Tsv("output.tsv"))
}
```

Run: using `scald.rb`

```
$ ./scald.rb --local project.scala --input in
```

Output: The output contains only contents of name field

```
Anil
Kumar
Srinivasa
Brad
```

7. **discard**: The API usage is

   ```
   pipe.discard(fields)
   ```

   Discard is the opposite of project function.

   Data: Consider a dataset of prices of commodities. First column is the commodity name, second column is its price.

   ```
   Nexus        23000
   Jersey       4000
   Football     2000
   Cricket-bat  1450
   ```

   Code: Discarding the product name, the result must contain only price.

   ```
   import com.twitter.scalding._

   class testJob(args: Args) extends Job(args) {
       val input = Tsv(args("input"), ('product, 'price))
           .discard('product)
           .write(Tsv("output.tsv"))
   }
   ```

   Run: using `scald.rb`

   ```
   $ ./scald.rb --local discard.scala --input in
   ```

   Output: The output contains only prices

   ```
   23000
   4000
   2000
   1450
   ```

8. **insert** : The API usage is

   ```
   pipe.insert(field, value)
   ```

   New entries are added to the fields.

   Data: Consider a product listing dataset of smartphones.

```
Samsung
Dell
Apple
Micromax
```

Code: Add a new field country of origin as "USA"

```
import com.twitter.scalding._

class testJob(args: Args) extends Job(args) {
    val input = Tsv(args("input"), ('product))
        .insert(('country), ("USA"))
        .write(Tsv("output.tsv"))
}
```

Run: using `scald.rb`

```
$ ./scripts/scald.rb --local insert.scala --input in
```

Output:

```
Samsung USA
Dell    USA
Apple   USA
Micromax    USA
```

9. **rename:** The API usage is

```
pipe.rename(oldFields -> newFields)
```

Data: Consider dataset of company names and company size

```
Samsung     40000
Dell        30000
Apple       50000
Micromax    10000
```

Code: Rename the fields from

```
('company, 'size) -> ('product, 'inventory).
```

```
import com.twitter.scalding._

class testJob(args: Args) extends Job(args) {
    val input = Tsv(args("input"), ('company, 'size))
        .rename(('company, 'size) ->
                        ('product, 'inventory))
        .project(('inventory, 'product))
        .write(Tsv("output.tsv"))
}
```

Run: using `scald.rb`

```
$ ./scald.rb --local rename.scala --input in
```

Output: The output is

```
40000    Samsung
30000    Dell
50000    Apple
10000    Micromax
```

10. **limit**: The API usage is

    ```
    pipe.limit(number)
    ```

    Limits the number of entries that flow through the pipe.

    Data: Consider the dataset of companies and revenues

    ```
    Samsung     40000
    Dell        30000
    Apple       50000
    Micromax    10000
    ```

    Code: Limit only to 2 companies that flow through the pipe.

    ```
    import com.twitter.scalding._

    class testJob(args: Args) extends Job(args) {
        val input = Tsv(args("input"), ('company, 'size))
            .limit(2)
            .write(Tsv("output.tsv"))
    }
    ```

    Run: using scald.rb

    ```
    $ ./scripts/scald.rb --local limit.scala --input in
    ```

    Output:

    ```
    Samsung 40000
    Dell    30000
    ```

11. **filter**: The API usage is

    ```
    pipe.filter(fields){function}
    ```

    Removes the rows for which the function evaluates to False.

    Data: Consider a dataset of animals and types.

    ```
    Crow       bird
    Lion       animal
    Sparrow    bird
    cat        animal
    Pigeon     bird
    ```

    Code: Filter out the animals of kind bird.

```
import com.twitter.scalding._

class testJob(args: Args) extends Job(args) {
    val input = Tsv(args("input"), ('animal, 'kind))
        .filter('kind) { kind : String => kind == "bird"}
        .write(Tsv("output.tsv"))
}
```

Run: using `scald.rb`

```
$ ./scripts/scald.rb --local filter.scala --input in
```

Output: Output of birds

```
Crow     bird
Sparrow bird
Pigeon   bird
```

12. **filterNot**: The API usage is

   `pipe.filterNot(fields) function`

   Works opposite to filter method.

   Data: Consider the dataset of animals.

```
Crow     bird
Lion     animal
Sparrow bird
Cat      animal
Pigeon   bird
```

   Code: Filter out animals that are not birds.

```
import com.twitter.scalding._

class testJob(args: Args) extends Job(args) {
    val input = Tsv(args("input"), ('animal, 'kind))
        .filterNot('kind) { kind : String => kind == "bird"}
        .write(Tsv("output.tsv"))
}
```

Run: using `scald.rb`

```
$ ./scripts/scald.rb --local filternot.scala --input in
```

Output: Animals that are not birds

```
Lion     animal
Cat      animal
```

13. **pack**: The API usage is

```
pipe.pack(Type)(fields -> object)
```

Using Java reflections multiple fields can be packed together into single object. Packing and unpacking are used to group and ungroup fields using objects.

Data: Consider a dataset of companies with name, size, and revenue (bn$).

```
Dell        40000    15
Facebook    23000    32
Google      47000    40
Apple       17000    34
```

Code: Pack the fields together as a Company object.

```scala
import com.twitter.scalding._

case class Company(companyID : String,
                   size : Long =0,
                   revenue : Int = 0)

class testJob(args: Args) extends Job(args) {

    val sampleinput = List(
        ("Dell",40000L,15),
        ("Facebook",23000L,32),
        ("Google",47000L,40),
        ("Apple",17000L,34))

    val input = IterableSource[(String, Long, Int)]
                (sampleinput, ('companyID, 'size, 'revenue))
        .pack[Company](
                ('companyID, 'size, 'revenue) -> 'Company)
        .write(Tsv("output.tsv"))
}
```

Run: using `scald.rb`

```
$ ./scripts/scald.rb --local pack.scala --input in
```

Output:

```
Dell      40000   15 Company(Dell,40000,15)
Facebook 23000   32 Company(Facebook,23000,32)
Google    47000   40 Company(Google,47000,40)
Apple     17000   34 Company(Apple,17000,34)
```

14. **unpack**: The API usage is

```
pipe.unpack(Type)(object -> fields)
```

Unpack the contents of an object into multiple fields

Data: Consider a packed dataset containing information about list of Companies with their names, size, and years since incorporation.

```
Company("Dell",40000,15)
Company("Facebook",23000,32)
Company("Google",47000,40)
Company("Apple",17000,34)
```

Code: Unpack the contents of the Company object into fields: `companyName`, `size`, `yearsOfInc`

```scala
import com.twitter.scalding._

case class Company(companyID : String,
                   size : Long =0,
                   revenue : Int = 0)

class testJob(args: Args) extends Job(args) {
   val sampleinput = List(
       Company("Dell",40000,15),
       Company("Facebook",23000,32),
       Company("Google",47000,40),
       Company("Apple",17000,34))

   val input = IterableSource[(Company)](sampleinput,
                                        ('company))
       .unpack[Company]('company ->
                   ('companyID, 'size, 'revenue))
       .write(Tsv("output.tsv"))
}
```

Run: using `scald.rb`

```
$ ./scald.rb --local unpack.scala
```

Output:

```
Company(Dell,40000,15)        Dell       40000   15
Company(Facebook,23000,32)    Facebook   23000   32
Company(Google,47000,40)      Google     47000   40
Company(Apple,17000,34)       Apple      17000   34
```

15. **groupBy**: The API usage is

```
pipe.groupBy(fields){group => <action>}
```

Groups data from pipe by the values of fields passed as arguments and then a set of operations are applied on these fields to create new set of fields.

Data: Consider the dataset of transactions on an e-commerce website. It consists of userId, and product bought.

```
1    camera
2    football
2    phone
1    sweater
1    shoes
```

```
1    shirt
3    laptop
```

Code: Find the number of items bought by each customer

The `size` grouping function is used to find the number of rows in each group.

```
import com.twitter.scalding._

class testJob(args: Args) extends Job(args) {
    val input = Tsv(args("input"), ('cust, 'product))
        .groupBy('cust){_.size}
        .write(Tsv("output.tsv"))
}
```

Run: using `scald.rb`

```
$ ./scald.rb --local groupby.scala --input in
```

Output: The output consists of customerId and number of purchases.

```
1    4
2    2
3    1
```

16. **groupAll**: The API usage is

```
pipe.groupAll{ group => <action> }
```

Create a single group of the entire pipe. This is useful in situations when you need a global variable.

Data: Consider the unordered phone book dataset

```
Kumar     657-737-8547
Anil      257-747-3527
Sunil     656-333-4542
Bob       617-730-8842
Rooney    125-679-0317
Falcao    957-717-3537
```

Code: Sort the dataset by names

```
import com.twitter.scalding._

class testJob(args: Args) extends Job(args) {
    val input = Tsv(args("input"), ('name, 'phone))
        .groupAll{_.sortBy('name)}
        .write(Tsv("output.tsv"))
}
```

Run: using `scald.rb`

```
$ ./scald.rb --local groupall.scala --input in
```

Output: Sorted phone book

```
Anil      257-747-3527
Bob       617-730-8842
Falcao    957-717-3537
Kumar     657-737-8547
Rooney    125-679-0317
Sunil     656-333-4542
```

17. **average**: The API usage is

    ```
    group.average(field)
    ```

    It is a grouping function used withing a grouping statement. It is used to find the mean of the group.

    Data: Consider a dataset of football players in a club with their ages.

    ```
    Anil         24
    Srinivasa    35
    Falcao       29
    Ronaldo      28
    Rooney       28
    Persie       31
    ```

    Code: Find the mean/average age of the club.

    ```
    import com.twitter.scalding._

    class testJob(args: Args) extends Job(args) {
        val input = Tsv(args("input"), ('player, 'age))
            .groupAll{ _.average('age) }
            .write(Tsv("output.tsv"))
    }
    ```

    Run: using scald.rb

    ```
    $ ./scald.rb --local average.scala --input in
    ```

    Output: Average age of the club is

    ```
    29.16
    ```

18. **sizeAveStdev**: The API usage is

    ```
    group.sizeAveStdev(field, fields)
    ```

    It calculates the count, average, and standard deviation over a field. The output fields are passed as arguments.

    Data: Consider the dataset of boys and girls with their ages

    ```
    Anil         24   boy
    Srinivasa    35   boy
    Falcao       29   boy
    ```

```
Betty         18   girl
Ronaldo       28   boy
Rooney        28   boy
Persie        31   boy
Veronica      26   girl
Sarah         24   girl
```

Code: Find the count, mean, and standard deviation for each group.

```
import com.twitter.scalding._

class testJob(args: Args) extends Job(args) {
    val input = Tsv(args("input"), ('name, 'age, 'sex))
        .groupBy('sex){ _.sizeAveStdev('age ->
                              ('count, 'mean, 'stdev))}
        .write(Tsv("output.tsv"))
}
```

Run: using `scald.rb`

```
$ ./scald.rb --local sizeavestdev.scala --input in
```

Output: Group-wise count, mean, and standard deviation, respectively.

```
boy   6    29.166666666666668    3.337497399083464
girl  3    22.666666666666668    3.39934634239519
```

19. **mkString**: The API usage is

```
group.mkString(field, joiner)
```

Takes a group and converts it into a string joined by joining string(e.g., comma, space, etc).

Data: Consider a chat dataset between two users they send to each other, words of data instead of a sentence.

```
George    meeting
Bob       today
George    the
George    is
Bob       is
George    when
Bob       meeting
George    Hi
Bob       Hi
```

Code: A hacker gains access to the chat server and needs to understand the chat by joining the chats of each person.

```
import com.twitter.scalding._

class testJob(args: Args) extends Job(args) {
    val input = Tsv(args("input"), ('name, 'chat))
```

```
        .groupBy('name){ _.mkString('chat, " ")}
        .write(Tsv("output.tsv"))
}
```

Run: using `scald.rb`

```
$ ./scald.rb --local mkstring.scala --input in
```

Output: Consists of group and joined words which can be randomly ordered.

```
Bob Hi meeting is today
George  Hi when is the meeting
```

20. **sum**: The API usage is

```
group.sum(field)
```

Calculates sum of all the group elements.

Data: Consider a dataset of purchases by a customer at different locations. It consists of shoppingLocation and purchaseAmount as columns.

```
Bangalore    12000
Delhi        3000
Bangalore    1000
Mumbai       2000
Delhi        30000
Bangalore    250
```

Code: Find the total amount spent at each location.

```
import com.twitter.scalding._

class testJob(args: Args) extends Job(args) {
    val input = Tsv(args("input"), ('location, 'amount))
         .groupBy('location){
             _.sum[Int]('amount -> 'total)
             }
         .write(Tsv("output.tsv"))
}
```

Run: using `scald.rb`

```
$ ./scald.rb --local sum.scala --input in
```

Output: Consists of shopping location and total amount.

```
Bangalore    13250
Delhi        33000
Mumbai       2000
```

21. **max, min**: The API usage is

```
group.max(field), group.min(field)
```

Computes the maximum element of the group, minimum element of the group, respectively.

Data: Consider a marks dataset of two students

```
Anil     45
Sunil    35
Anil     23
Sunil    22
Anil     56
Sunil    57
```

Code: Find the maximum and minimum marks for each student.

```scala
import com.twitter.scalding._

class testJob(args: Args) extends Job(args) {
    val input = Tsv(args("input"), ('student, 'marks))
        .groupBy('student) {
            _.max('marks -> 'max)
            .min('marks -> 'min)
            }
        .write(Tsv("output.tsv"))
}
```

Run: using `scald.rb`

```
$ ./scald.rb --local maxmin.scala --input in
```

Output: Consists of student, maximum marks, and minimum marks

```
Anil     56    23
Sunil    57    22
```

22. **count** : The API usage is

`group.count(field){function}`

Counts the number of rows in the group that satisfies a condition defined by the function.

Data: Consider the dataset of run scored by a team during a match

```
Anil      45
Jadeja    35
Dhoni     23
Rohit     43
Kumar     56
Dravid    57
```

Code: Find the number of players who have scored more than 40 runs

```scala
import com.twitter.scalding._

class testJob(args: Args) extends Job(args) {
```

```scala
val input = Tsv(args("input"), ('player, 'runs))
    .groupAll{
        _.count(('player, 'runs) -> 'c){
            x : (String, Int) =>
            val (player, runs) = x
            (runs > 40)
        }
    }
    .write(Tsv("output.tsv"))
}
```

Run: using `scald.rb`

```
$ ./scald.rb --local count.scala --input in
```

Output: Number of players with more than 40 runs are

4

23. **sortBy:** The API usage is

`group.sortBy(fields)`

Sort the group before spilling to storage

Data: Consider a dataset of population of countries. It contains country name and population.

```
India        1,248,820,000
China        1,366,540,000
USA          318,679,000
Indonesia    252,164,800
Brazil       203,097,000
```

Code: Sort the dataset by country name

```scala
import com.twitter.scalding._

class testJob(args: Args) extends Job(args) {
    val input = Tsv(args("input"),
                    ('country, 'population))
        .groupAll{
            _.sortBy('country)
        }
        .write(Tsv("output.tsv"))
}
```

Run: using `scald.rb`

```
$ ./scald.rb --local sortby.scala --input in
```

Output:

```
Brazil       203,097,000
China        1,366,540,000
```

```
India        1,248,820,000
Indonesia    252,164,800
USA          318,679,000
```

Note: **Replace** `sortBy('country)` with `sortBy('country).` `reverse` to sort in reverse order. Try it.

24. **reduce**: The API usage is

```
group.reduce(field){function}
```

Performs a reduce operation on a group. If the reduce operation is associative it can be done on the map side instead of on the reduce side, similar to a combiner.

Data: Consider a dataset of purchases in different locations.

```
Bangalore    10000
Delhi        12000
Bangalore     3000
Delhi         2000
Delhi         1500
Bangalore     3000
```

Code: Find the total expenditure per location.

```
import com.twitter.scalding._

class testJob(args: Args) extends Job(args) {
    val input = Tsv(args("input"),
                          ('location, 'amount))
            .groupBy('location){
                _.reduce('amount -> 'total){
                    (temp : Int, amount : Int) =>
                           temp + amount
                }
            }
            .write(Tsv("output.tsv"))
}
```

Run: using `scald.rb`

```
$ ./scald.rb --local reduce.scala --input in
```

Output: Total cost per location is

```
Bangalore    16000
Delhi        15500
```

25. **foldLeft**: The API usage is

```
group.foldLeft(field){function}
```

Fold action is similar to reduce operation but takes place strictly on the reduce side. The operation need not be associative. The operation iterates from the left side of a list in the group.

Data: Consider a product dataset and purchase status of customers.

```
Sunil    ProductA    false
Sunil    ProductB    true
Anil     ProductA    false
Anil     ProductB    false
Kumar    ProductA    true
Kumar    ProductB    true
```

Code: Check if the customers have bought at least one of the two products.

```scala
import com.twitter.scalding._

class testJob(args: Args) extends Job(args) {
    val input = Tsv(args("input"),
                        ('customer, 'product, 'bought))
            .groupBy('customer) {
                _.foldLeft('bought -> 'bought)(false) {
                    (prev : Boolean, current : Boolean) =>
                        prev || current
                }
            }
            .write(Tsv("output.tsv"))
}
```

Run: using `scald.rb`

```
$ ./scald.rb --local foldleft.scala --input in
```

Output: Consists of customer and product bought status

```
Anil     false
Kumar    true
Sunil    true
```

26. **take:** The API usage is

```
group.take(number)
```

Take the first `number` of items from the group.

Data: Consider a dataset of students with marks out of 200

```
Anil        110
Bob          98
Robert      197
Sarah       112
Betty        54
Veronica    165
Simon       123
Rooney       99
```

Code: Calculate the student rank list and select the top three students with their marks

```
import com.twitter.scalding._

class testJob(args: Args) extends Job(args) {
    val input = Tsv(args("input"), ('student, 'marks))
        .read
        .map('marks -> 'marksInt){ x : Int => x}
        .discard('marks)
        .groupAll{
            _.sortBy('marksInt).reverse
            .take(3)
        }
        .write(Tsv("output.tsv"))
}
```

Run: using `scald.rb`

```
$ ./scald.rb --local take.scala --input in
```

Output: The top 3 students in the class are

```
Robert    197
Veronica  165
Simon     123
```

Note: Replace

`.take(3)` with

`.takeWhile('marksInt){ x : Int => x > 100}`

in the code above and execute it similarly. It selects all the students with marks more than 100

Output: Students with marks more than 100 are

```
Robert    197
Veronica  165
Simon     123
Sarah     112
Anil      110
```

27. **drop:** The API usage is

`group.drop(number)`

Similar to `take`, it drops the first `number` of elements of the group.

Data: Consider a dataset of unique unordered list of student roll numbers

```
2
5
7
```

```
8
6
9
4
1
10
3
```

Code: Select students with roll numbers greater than 3

```
import com.twitter.scalding._

class testJob(args: Args) extends Job(args) {
    val input = Tsv(args("input"), ('numbers))
        .read
        .mapTo('numbers -> 'numbersInt){
                                    x : Int => x}
        .groupAll{
            _.sortBy('numbersInt)
            .drop(3)
        }
        .write(Tsv("output.tsv"))
}
```

Run: using scald.rb

```
$ ./scald.rb --local drop.scala --input in
```

Output: Roll numbers greater than 3 are

```
4
5
6
7
8
9
10
```

28. **sortedWithTake**: The API usage is

```
group.sortWithTake( currentFields -> newField, take
Number)
```

This is similar to sorting with a function and then taking k items. It is efficient compared to performing a total sort followed by a take as the sort in this case is done on the mapper.

Data: Consider the product listing in an online marketplace, such that it has categoryKey, productID, and rating as entries.

```
"a"      2        3.0
"a"      3        3.0
"a"      1        3.5
"b"      1        6.0
```

```
"b"      2        5.0
"b"      3        4.0
"b"      4        3.0
"b"      5        2.0
"b"      6        1.0
```

Code: Group and sort the dataset such that the products are sorted against the ratings and then their IDs if they have same rating.

```scala
import com.twitter.scalding._

class testJob(args: Args) extends Job(args) {
    val input = Tsv(args("input"),
                         ('key, 'product_id, 'rating))
        .read
        .map(('key, 'product_id, 'rating) ->
                   ('key, 'product_id, 'rating)){
            x : (String, Int, Double) =>
            val (key, product_id, rating) = x
            (key, product_id, rating)
        }
        .groupBy('key) {
          _.sortWithTake[(Int, Double)](((('product_id,
                                           'rating),
                                          'top_products),
                                          5) {
            (product_0: (Int, Double),
             product_1: (Int, Double)) =>
              if (product_0._2 == product_1._2) {
                  product_0._1 < product_1._1
              }
              else {
                  product_0._2 > product_1._2
              }
          }
        }
        .write(Tsv("output.tsv"))
}
```

Run: using `scald.rb`

```
$ ./scald.rb --local sortWithTake.scala --input in
```

Output: The sorted category listings are as follows: categoryID, sorted list. Category "b" has only 5 elements since k=5.

```
"a" List((1,3.5), (2,3.0), (3,3.0))
"b" List((1,6.0), (2,5.0), (3,4.0), (4,3.0), (5,2.0))
```

29. **pivot** : The API usage is

```
group.pivot((fields, fields) -> (pivoting values
fields), defaultValue)
```

Pivoting happens over two columns where repeated elements can be grouped such that the corresponding values in the other columns are arranged properly.

Data: Consider a dataset of university with course wise student strength. The dataset consists of course, semester, and strength as columns

```
computer     sem1     120
computer     sem2     200
electrical   sem3     150
electrical   sem2     150
computer     sem3     140
electrical   sem1     150
```

Code: Pivot the dataset on semesters

```scala
import com.twitter.scalding._

class testJob(args: Args) extends Job(args) {
    val input = Tsv(args("input"),
                    ('course, 'semester, 'students))
        .read
        .groupBy('course) {
           _.pivot(('semester, 'students) ->
                              ('sem1, 'sem2, 'sem3), 0)
        }
        .write(Tsv("output.tsv"))
}
```

Run: using `scald.rb`

```
$ ./scald.rb --local pivot.scala --input in
```

Output: Pivoted over courses.

```
             sem1    sem2    sem3
computer     120     200     140
electrical   150     150     150
```

unpivot : The API usage is

```
pipe.unpivot(pivoted values) ->          (comma separted
fields)
```

Add the following code to the previous example.

```scala
input
   .unpivot(('sem1, 'sem2, 'sem3)-> (('semester, 'students)))
   .write(Tsv("unpivot.tsv"))
```

output: Unpivoted values look like the original input

```
computer     sem1     120
computer     sem2     200
computer     sem3     140
electrical   sem1     150
```

```
electrical    sem2    150
electrical    sem3    150
```

Join Operations: They merge two pipes on set of keys, similar to SQL join operations. Cascading implements join as CoGroup operations.

30. **joinWithSmaller**: The API usage is

    `pipe1.joinWithSmaller(fields, pipe2)` This is preferred when right pipe, that is pipe2, is relatively smaller compared to pipe1.

    Data: Consider dataset of student performances in two subjects. The dataset is arranged as studentID and marks columns.

```
342       99
213       76
244       65
352       96
546       34
446       57

352       34
546       96
446       47
342       76
213       99
244       56
```

Code: Join the student performance over studentID.

```
import com.twitter.scalding._

class testJob(args: Args) extends Job(args) {
    val pipe1 = Tsv(args("in1"),('studid1, 'sub1))
        .read
    val pipe2 = Tsv(args("in2"), ('studid2, 'sub2))
        .read

    pipe1.joinWithSmaller('studid1 ->
                                    'studid2, pipe2)
        .discard('studid2)
        .write(Tsv("output.tsv"))

}
```

Run: using `scald.rb`

```
$ ./scald.rb --local joinwithsmaller.scala --in1 in --in2 in2
```

Output: After join the columns are studentID, subject1, and subject2.

```
213      76      99
244      65      56
342      99      76
352      96      34
446      57      47
546      34      96
```

joinWithLarger: The API usage is

```
pipe1.joinWithLarger(fields, pipe2)
```

This API is used when the size of pipe2 is much larger than pipe1. In the code above, the only change is to

Replace

```
pipe1.joinWithSmaller('studid1 -> 'studid2, pipe2)
```

with

```
pipe1.joinWithLarger('studid1 -> 'studid2, pipe2)
```

joinWithTiny: The API usage is

```
pipe1.joinWithTiny(fields, pipe2)
```

It is special join operation where the join happens on the map side. It does not move the left pipe to the reducers, instead the entire right is replicated by the mappers. This can be used when

```
# rows in pipe1 > # of mappers * # rows in pipe2
```

In the previous code example, replace

```
pipe1.joinWithSmaller('studid1 -> 'studid2, pipe2)
```

with

```
pipe1.joinWithTiny('studid1 -> 'studid2, pipe2)
```

There are several joining modes, as is common in SQL queries. The `joiner` argument in the join operations requires

```
import cascading.pipe.joiner._
```

Scalding provides the following joiner classes:

- **LeftJoin**: It keeps all the rows of the left pipe and matches the entries of the right pipe.

  The API usage is

  ```
  pipe1.joinWithSmaller('studid1 -> 'studid2, pipe2,
  joiner = new LeftJoin)
  ```

- **RightJoin**: It keeps all the entries of the right pipe and attaches the entries of left pipe

  The API usage is

  ```
  pipe1.joinWithSmaller('studid1 -> 'studid2, pipe2,
  joiner = new RightJoin)
  ```

- **OuterJoin**: It keeps the entries of both the pipes

  The API usage is

  ```
  pipe1.joinWithSmaller('studid1 -> 'studid2, pipe2,
  joiner = new OuterJoin)
  ```

  Note: If there are entries that do not match, the empty fields contain `null`.

31. **crossWithTiny**: The API usage is

    ```
    pipe1.crossWithTiny(pipe2)
    ```

    Peforms a cross product of the fields. It becomes computationally intensive when the right pipe increases in size.

    Data: Consider a dataset of books such that there is a list of books with title and another dataset with common location data.

    ```
    chemistry
    physics
    biology
    math

    Bangalore
    Delhi
    ```

    Code: Add another column to the dataset with the library locations that are common to all the books.

    ```
    import com.twitter.scalding._

    class testJob(args: Args) extends Job(args) {
        val pipe1 = Tsv(args("in1"),('bookid, 'title))
            .read
        val pipe2 = Tsv(args("in2"), ('location))
            .read

        pipe1.crossWithTiny(pipe2).write(Tsv("output.tsv"))
    }
    ```

    Run: using `scald.rb`

    ```
    $ ./scald.rb --local crosswithtiny.scala --in1 in --in2 in2
    ```

Output: Contains a list of book titles and available locations

```
chemistry Bangalore
chemistry Delhi
physics   Bangalore
physics   Delhi
biology   Bangalore
biology   Delhi
math      Bangalore
math      Delhi
```

32. **dot**: The API usage is

```
groupBy('x) { _.dot('y,'z, 'y_dot_z) }
```

Performs a dot product on two fields y and z results in dot product field. Given a group of two rows the $y_1 dot z_1 + y_2 dot z_2$.

Data: Consider a dataset of changing rectangle shapes with varying length and width

```
rectangle1    2      10
rectangle2    12     20
rectangle1    14     17
rectangle2    6      6
rectangle5    17     25
rectangle6    9      18
```

Code: Calculate the dot product of these rectangle shapes

```
import com.twitter.scalding._

class testJob(args: Args) extends Job(args) {
    val pipe1 = Tsv(args("in1"),('shape, 'len, 'wid))
        .read
        .groupBy('shape){
            _.dot[Int]('len, 'wid, 'len_dot_wid)
        }
        .write(Tsv("output.tsv"))
}
```

Run: using scald.rb

```
$ ./scald.rb --local dot.scala --in1 in
```

Output: After dot product

```
rectangle1    258
rectangle2    276
rectangle5    425
rectangle6    162
```

33. **normalize**: The API usage is

```
pipe.normalize(fields)
```

Performs normalization over a column, which is equivalent to performing

$$norm(x_i) = \frac{x_i}{\sum x_i} \quad \forall x_i \in S.$$

Data: Consider a dataset of player ratings of a league

```
player1 90
player2 45
player3 76
player4 75
player5 54
player6 87
```

Code: Normalize the player ratings

```scala
import com.twitter.scalding._

class testJob(args: Args) extends Job(args) {
    val pipe1 = Tsv(args("in1"),('player, 'rating))
        .read
        .normalize('rating)
        .write(Tsv("output.tsv"))
}
```

Run: using `scald.rb`

```
$ ./scald.rb --local normalize.scala --in1 in
```

Output: The normalized player ratings

```
player1  0.2107728337236534
player2  0.1053864168618267
player3  0.17798594847775176
player4  0.1756440281030445
player5  0.12646370023419204
player6  0.20374707259953162
```

34. **addTrap**: The API usage is

```
pipe.addTrap(location)
```

Works as a trigger that captures any exceptions that are thrown in the pipe.

Data: Consider a simple dataset that contains two interger columns

```
10      1
4       3
5       8
4       3
1       0
2       2
```

Code: Perform a simple divide operation on column1 over column2. Since the data has 0 in column2, we can expect a divide-by-zero exception.

```
import com.twitter.scalding._

class testJob(args: Args) extends Job(args) {
    val pipe1 = Tsv(args("in1"),('x, 'y))
        .read
        .map(('x, 'y)-> 'div){
            x : (Int, Int) => x._1 / x._2
        }
        .addTrap(Tsv("output-trap.tsv"))
        .write(Tsv("output.tsv"))
}
```

Run: using `scald.rb`

```
$ ./scald.rb --local addtrap.scala --in1 in
```

Output: After simple division, the divide-by-zero exception has been caught

output.tsv looks like this

```
10   1    10
4    3    1
5    8    0
4    3    1
2    2    1
```

output-trap.tsv that captures the exceptions looks like this

```
1    0
```

35. **sample:** The API usage is

```
pipe.sample(percentage)
```

Select a sample of the pipe, the percentage argument ranges between 0.0 for 0% and 1.0 for 100%.

Data: Consider a dataset of students of large university

```
student1     196
student2     285
student3     375
student4     464
student5     553
student6     642
student7     731
student8     821
student9     930
student10    719
```

Code: Select the top 80 % of the class.

```
import com.twitter.scalding._

class testJob(args: Args) extends Job(args) {
    val pipe1 = Tsv(args("in1"),('student, 'marks))
        .read
        .map('marks -> 'marksInt){
            x : Int => x
        }
        .discard('marks)
        .groupAll{
            _.sortBy('marksInt).reverse
        }
        .sample(0.8)
        .debug
        .write(Tsv("output.tsv"))
}
```

Run: using `scald.rb`

```
$ ./scald.rb --local sample.scala --in1 in
```

Output: Top 80 % of the class looks like

```
student9      930
student8      821
student7      731
student10     719
student6      642
student5      553
student4      464
student1      196
```

36. **Combinatorics** Permutations and combinations are two major aspects of combinatorics. The APIs are listed in the package `com.twitter.scalding.mathematics` in class `Combinatorics`

Code: Find the permutation $^6P_3$ and combination $^5C_2$

The code below generates all the permutations and combinations as defined.

```
import com.twitter.scalding._
import com.twitter.scalding.mathematics._
class testJob(args: Args) extends Job(args) {
    val c = Combinatorics
    c.permutations(6, 3).write(Tsv("permutations.txt"))
    c.combinations(5, 2).write(Tsv("combinations.txt"))
}
```

Run: using `scald.rb`

```
$ ./scald.rb --local combinatorics.scala
```

Output: Permutations

```
5          6          1
5          6          2
5          6          3
5          6          4
2          3          1
2          3          4
2          3          5
```

. . .

Output: Combinations

```
1          2
1          3
1          4
1          5
4          5
3          4
3          5
2          3
2          4
2          5
```

Code: Determine the permutation and combination by determining the size of the list generated.

```
import com.twitter.scalding._
import com.twitter.scalding.mathematics._
class testJob(args: Args) extends Job(args) {
    val perm = TextLine("permutations.txt")
        .groupAll{
            _.size
        }
        .debug
        .write(TextLine("poutput.tsv"))

    val comb = TextLine("combinations.txt")
        .groupAll{
            _.size
        }
        .debug
        .write(TextLine("coutput.tsv"))
}
```

Run: using scald.rb

```
$ ./scald.rb --local combinatorics_checker.scala
```

Output: Counting the size of the number entries in the list gives the value of

```
120
10
```

4.2 Spark Programming Guide

An overview of Spark application consists of a `driver` program that executes the Spark main that starts various operations on a cluster in parallel. All these operations are performed on top of data distribution abstraction called Resilient Distributed Datasets (RDDs), which is a collection of elements that are spread across the cluster as partitions with high fault tolerance. Spark offers the RDD to be stored in-memory allowing it to efficiently reuse the data across several parallel operations.

Another important abstraction is `shared variables` which can be used across multiple parallel operations. Normally when you run Spark or any such distributed computing application, all the applications related variables functions are shipped to different nodes. Sometimes a shared variable needs to be shared across tasks. Spark provides two types of shared variables `broadcast`, which can used to cache a value in-memory and `accumulators`, which can used as counters and sums.

Broadcast Variables

They are used to store read-only values in-memory instead of copying it across all the nodes in the cluster. For instance, they can be used to store a large input dataset that can be distributed efficiently using broadcasting algorithms that reduce communication costs.

They can be initialized using `SparkContext.broadcast(v)`, where `v` is the value that is shared. The value can be accessed by the `value` attribute.

```
scala> val broadcaster = sc.broadcast(Array(5,6,7,8))
broadcaster: org.apache.spark.broadcast.Broadcast[Array[Int]]
    = Broadcast(0)

scala> broadcaster.value
res0: Array[Int] = Array(5, 6, 7, 8)
```

Accumulators

These variables use typically function as counters and sums. They can only be added to an associative operation and thus are very efficient during parallel operations. Sparks supports `accumulators` or type `Int` and `Double`.

They are similar to broadcast variables in initialization, using `SparkContext.accumulator(v)`. They can be added to use a `+=` operator. The value of an accumulator can be accessed using `value` attribute.

```
scala> val accum = sc.accumulator(0)
accum: org.apache.spark.Accumulator[Int] = 0

scala> sc.parallelize(Array(1, 2, 3, 4)).foreach(x => accum
    += x)
```

```
scala> accum.value
res2: Int = 10
```

Input | Output

RDDs can be created from any file stored in storage systems like HDFS, local file system, Amazon S3, etc. Spark also supports textFile, SequenceFile, and Hadoop InputFormat.

SparkContext provides a textFile method that takes a URI of the form hdfs://, s3n://, etc.

```
scala> val distFile = sc.textFile("README.md")
distFile: org.apache.spark.rdd.RDD[String] = MappedRDD[1] at
    textFile at <console>:12

scala> distFile.count()
14/09/05 01:24:44 INFO SparkContext: Starting job: count at <
    console>:15
14/09/05 01:24:44 INFO DAGScheduler: Got job 2 (count at <
    console>:15) with 2 output partitions (allowLocal=false)
14/09/05 01:24:44 INFO DAGScheduler: Final stage: Stage 2(count
    at <console>:15)
14/09/05 01:24:44 INFO DAGScheduler: Parents of final stage:
    List()
14/09/05 01:24:44 INFO DAGScheduler: Missing parents: List()
...
14/09/05 01:24:44 INFO DAGScheduler: Stage 2 (count at <console
    >:15) finished in 0.011 s
14/09/05 01:24:44 INFO SparkContext: Job finished: count at <
    console>:15, took 0.015264842 s
res3: Long = 127
```

Once distFile RDD is created, actions and transformations can be applied to it. The distFile is an RDD of a text file, the number of lines in the file can be calculated by using count() action.

Note: The textFile method takes another argument to control the number of file slices. By default the number of slices is one for each block, this can be increased to have fewer slices than blocks.

Another input source can be SparkContext's sequenceFile[K, V], where K and V are the types of keys and values in the file, respectively. They are subclasses of Hadoop Writable class, like IntWritable and Text.

Parallelized Collections

Scala collections can be parallelized using SparkContext's `parallelize` method. The elements of the collection are replicated to form a RDD that can be operated on in parallel.

```scala
scala> val data = Array(1, 2, 3, 4, 5)
data: Array[Int] = Array(1, 2, 3, 4, 5)

scala> val distData = sc.parallelize(data)
distData: org.apache.spark.rdd.RDD[Int] = ParallelCollectionRDD
    [0] at parallelize at <console>:14

scala> distData.reduce(_ + _)
...
14/09/05 02:00:44 INFO DAGScheduler: Stage 3 (reduce at <console
    >:17) finished in 0.019 s
14/09/05 02:00:44 INFO TaskSchedulerImpl: Removed TaskSet 3.0,
    whose tasks have all completed, from pool
14/09/05 02:00:44 INFO SparkContext: Job finished: reduce at <
    console>:17, took 0.032209914 s
res5: Int = 6
```

Once the RDD `distData` has been created, operations like `reduce` can be applied. Here `distData.reduce(_ + _)` adds the contents of the array.

The `parallelize` method also takes another argument that indicates the number of slices. Spark runs one task per slice.

RDD Persistence

Among Spark's most important features are its ability to `persist` to any storage and `cache` in-memory. Persist operation on RDD stores its slices in-memory and reuses them when required. This in-memory data availability allows actions to be faster atleast by 10x. Caching is used to build most iterative algorithms.

`persist` and `cache` actions can be used to persist an RDD. These actions are fault-tolerant, in the event of a failure the lost partition will be recomputed from the parent dataset following the transformation history that is stored.

RDD can be persisted onto the disk at different levels. These levels are instances of `org.apache.spark.storage.StorageLevel`. `cache` is used as a shorthand for `StorageLevel.MEMORY_ONLY`. Other storage levels are described in Table 4.3.

Table 4.3: Storage level for persisting [5]

Storage Level	Meaning
MEMORY_ONLY	Store the objects in RDD. They are not serialized. If they do not fit memory, they are recomputed when needed. Default.
MEMORY_AND_DISK	Stores the RDD in the JVM. If the RDD does not fit, those partitions are spilled to the disk and read from when needed.
MEMORY_ONLY_SER	They are more space efficient as they are serialized. They are stored in memory. They are more computationally intensive to read.
MEMORY_AND_DISK_SER	As in MEMORY_ONLY_SER, but spill partitions that do not fit in memory to disk instead of recomputing them on the fly each time they are needed.
DISK_ONLY	Store the RDD partitions only on disk.
MEMORY_ONLY_2, MEMORY_AND_DISK_2, etc.	Same as the levels above, but replicate each partition on two cluster nodes.

Spark's storage level options provide different choices for different trade-offs between memory and CPU. To make a choice some important details need to be understood.

- Use MEMORY_ONLY when the RDDs fit into memory. This is highly CPU efficient and allows the operations to run as fast as possible.

- Next option, use MEMORY_ONLY_SER with a fast serialization library to make the objects much more space-efficient, but still reasonably fast to access.

- Do not use the secondary storage unless the computations are expensive and the dataset is large. Recomputing is faster than reading from disk.

- Fault tolerance while using the disk can be ensured by using replicated storage levels. All storage levels provide fault tolerance by recomputing lost partitions but the replicated ones let you run the computation without waiting for lost partition.

RDD Operations

The operations on RDDs are of two types: *Transformations* and *Actions*. Transformations create a new dataset from a current, while actions perform a computation on the dataset and return a value to the driver program. For example, a map transformation passes each element of dataset to a function and returns another dataset with processed elements. The reduce action performs an aggregation operation on all the elements of the dataset and return a final result to the driver program.There are many such actions and transformations, some of them are explained below with example code.

Actions and Transformation

Start the Spark for all the examples shown hereafter. The SparkContext is initialized and is used as `sc`.

Spark Shell: Start the shell by executing `spark-shell` script

```
$ ./bin/spark-shell
...
14/09/05 09:40:22 INFO SparkILoop: Created spark context..
Spark context available as sc.

scala>
```

1. **`parallelize`**: The API usage is

   ```
   sc.parallelize(sequence, numSlices)
   ```

   Parallelize acts lazily over the sequence of data. If the sequence is mutable then any changes to it before the first action on the RDD will be reflected in the RDD.

   `return:` **RDD** of type `ParallelCollectionRDD`.

   Note: `makeRDD` is another function definition to parallelize the data. In its implementation, it calls the above `parallelize` method.

   Spark Shell: Create an RDD for an array sequence

   ```
   scala> val data = sc.parallelize(Array(1,2,3,4))
   data: org.apache.spark.rdd.RDD[Int] = ParallelCollectionRDD
       [0] at parallelize at <console>:12

   scala> data.collect()
   14/09/05 09:42:08 INFO SparkContext: Starting job: collect at
       <console>:15
   ...
   res0: Array[Int] = Array(1, 2, 3, 4)
   ```

2. **`map`**: The API usage is

   ```
   RDD.map(function)
   ```

   Returns a new RDD by passing each element of the RDD through a `function`

   Spark Shell: Create an RDD of array of five integers and map each element to multiply by 2

   ```
   scala> val data = sc.makeRDD(Array(1,2,3,4))
   data: org.apache.spark.rdd.RDD[Int] = ParallelCollectionRDD
       [0] at makeRDD at <console>:12

   scala> data.map(x => x*2).collect()
   14/09/05 12:01:57 INFO SparkContext: Starting job: collect at
       <console>:15
   ```

```
...
14/09/05 12:01:57 INFO SparkContext: Job finished: collect at
    <console>:15, took 0.164223172 s
res0: Array[Int] = Array(2, 4, 6, 8)
```

3. **filter**: The API usage is

   ```
   RDD.filter(function)
   ```

   Creates an RDD by selecting the elements whose return value is True on application of function.

   Spark Shell: Filter elements that are greater than 4 from an Array(1,..,10).

   ```
   scala> val data = sc.makeRDD(Array(1,2,3,4,5,6,7,8,9,10))
   data: org.apache.spark.rdd.RDD[Int] = ParallelCollectionRDD
       [0] at makeRDD at <console>:12

   scala> val fildata = data.filter(x => x > 4)
   fildata: org.apache.spark.rdd.RDD[Int] = FilteredRDD[2] at
       filter at <console>:14

   scala> fildata.collect()
   14/09/05 12:08:30 INFO SparkContext: Starting job: collect at
       <console>:17

   14/09/05 12:08:30 INFO SparkContext: Job finished: collect at
       <console>:17, took 0.024437998 s
   res3: Array[Int] = Array(5, 6, 7, 8, 9, 10)
   ```

4. **flatMap**: The API usage is

   ```
   RDD.flatMap(function)
   ```

   Returns a new RDD after applying a function to all the elements and then flattening the results.

   Spark Shell: For each number x of an Array generate a sequence starting from 1 to x.

   ```
   scala> val data = sc.makeRDD(Array(1,2,3,4))
   data: org.apache.spark.rdd.RDD[Int] = ParallelCollectionRDD
       [0] at makeRDD at <console>:12

   scala> val newdata = data.flatMap(x => 1 to x)
   newdata: org.apache.spark.rdd.RDD[Int] = FlatMappedRDD[1] at
       flatMap at <console>:14

   scala> newdata.collect()
   14/09/05 12:20:00 INFO SparkContext: Starting job: collect at
       <console>:17

   ...
   ```

```
14/09/05 12:20:01 INFO SparkContext: Job finished: collect at
    <console>:17, took 0.175433931 s
res0: Array[Int] = Array(1, 1, 2, 1, 2, 3, 1, 2, 3, 4)
```

5. **mapPartitions**: The API usage is

```
RDD.mapPartitions(function)
```

Returns a new RDD by applying `function` on each partition.

Spark Shell: Add all the elements in each partition and return the partition sums. Here the contents of the partitions can be determined by using the `glom` feature.

```
scala> val data = sc.makeRDD(Array(1,2,3,4,5,6))
data: org.apache.spark.rdd.RDD[Int] = ParallelCollectionRDD
    [18] at makeRDD at <console>:12

scala> data.glom().collect()
14/09/05 13:41:15 INFO SparkContext: Starting job: collect at
    <console>:15
...
14/09/05 13:41:15 INFO SparkContext: Job finished: collect at
    <console>:15, took 0.023524513 s
res3: Array[Array[Int]] = Array(Array(1), Array(2, 3), Array
    (4), Array(5, 6))

scala> data.mapPartitions(iter => Iterator(iter.reduce(_ + _)
    )).collect()
14/09/05 13:42:16 INFO SparkContext: Starting job: collect at
    <console>:15
...
14/09/05 13:42:16 INFO SparkContext: Job finished: collect at
    <console>:15, took 0.026995335 s
res7: Array[Int] = Array(1, 5, 4, 11)
```

6. **mapPartitionsWithIndex**: The API usage is

```
RDD.mapPartitionsWithIndex(function)
```

Returns a new RDD by applying a `function` on the elements of each partition, while tracking the index of each partition.

Spark Shell: Add the elements of each partition and return the sum for each partition.

```
scala> val data = sc.makeRDD(Array(1,2,3,4,5,6))
data: org.apache.spark.rdd.RDD[Int] = ParallelCollectionRDD
    [0] at makeRDD at <console>:12

scala> data.mapPartitionsWithIndex{
    | case(partition, iter) => Iterator((partition, iter.
        reduce(_+_)))
    | }.collect()
```

```
14/09/05 13:49:27 INFO SparkContext: Starting job: collect at
    <console>:17
...
14/09/05 13:49:27 INFO SparkContext: Job finished: collect at
    <console>:17, took 0.166275988 s
res1: Array[(Int, Int)] = Array((0,1), (1,5), (2,4), (3,11))
```

7. **distinct**: The API usage is

RDD.distinct() with optional numPartitions argument.

Returns a new RDD containing the unique elements.

Spark Shell: Find the distinct elements from a collection of items with more than one occurrence.

```
scala> val data = sc.makeRDD(Array
    (1,1,1,1,2,2,2,3,4,5,6,7,8,8,8,8,9))
data: org.apache.spark.rdd.RDD[Int] = ParallelCollectionRDD
    [0] at makeRDD at <console>:12

scala> val newdata = data.distinct()
newdata: org.apache.spark.rdd.RDD[Int] = MappedRDD[5] at
    distinct at <console>:14

scala> newdata.collect()
14/09/05 12:27:33 INFO SparkContext: Starting job: collect at
    <console>:17
...
14/09/05 12:27:33 INFO SparkContext: Job finished: collect at
    <console>:17, took 0.377903629 s
res0: Array[Int] = Array(4, 8, 1, 9, 5, 6, 2, 3, 7)
```

8. **groupByKey**: The API usage is

RDD.groupByKey()

Returns an RDD by grouping the values for each key in the RDD into a single sequence.

Spark Shell: Create a dictionary of key and values from an array of repeated (key, value) tuples.

```
scala> val data = sc.makeRDD(Array((1,"a"),(2,"b"),(1,"c")
    ,(2,"d")))
data: org.apache.spark.rdd.RDD[(Int, String)] =
    ParallelCollectionRDD[0] at makeRDD at <console>:12

scala> data.groupByKey().collect()
14/09/05 13:29:45 INFO SparkContext: Starting job: collect at
    <console>:15
...
```

```
14/09/05 13:29:45 INFO SparkContext: Job finished: collect at
    <console>:15, took 0.370931499 s
res0: Array[(Int, Iterable[String])] = Array((1,ArrayBuffer(a
    , c)), (2,ArrayBuffer(b, d)))
```

9. **reduceByKey**: The API usage is

```
RDD.reduceByKey(function)
```

Returns an RDD by merging the values for each key using an associative reduce function.

Spark Shell: Find the sum of all the values associated with each key.

```
scala> val data = sc.makeRDD(Array((1,1),(2,3),(1,2),(2,1)
    ,(1,4),(2,3)))
data: org.apache.spark.rdd.RDD[(Int, Int)] =
    ParallelCollectionRDD[2] at makeRDD at <console>:12

scala> data.reduceByKey(_+_).collect()
14/09/05 13:53:55 INFO SparkContext: Starting job: collect at
    <console>:15
...
14/09/05 13:53:55 INFO SparkContext: Job finished: collect at
    <console>:15, took 0.203548579 s
res2: Array[(Int, Int)] = Array((1,7), (2,7))
```

10. **sortByKey**: The API usage is

```
RDD.sortByKey()
```

Returns an RDD with sorted elements by the key.

Spark Shell: Sort the array of (k,v) pairs by the key

```
scala> val data = sc.makeRDD(Array((1,1),(2,3),(1,2),(2,1)
    ,(1,4),(2,3)))
data: org.apache.spark.rdd.RDD[(Int, Int)] =
    ParallelCollectionRDD[19] at makeRDD at <console>:16

scala> data.sortByKey().collect()
14/09/05 14:04:09 INFO SparkContext: Starting job: sortByKey
    at <console>:19
...
14/09/05 14:04:09 INFO SparkContext: Job finished: collect at
    <console>:19, took 0.051977158 s
res11: Array[(Int, Int)] = Array((1,1), (1,2), (1,4), (2,3),
    (2,1), (2,3))
```

11. **join**: The API usage is

```
RDD1.join(RDD2)
```

Returns a new RDD containing all the pairs of elements with matching keys between two RDDs. The elements will be returned as (k, (v1, v2)) tuple where (k, v1) is present in one RDD and (k, v2) is present in the other.

Spark Shell:

```
scala> val data1 = sc.parallelize(Array((1, 23), (1, 25), (2,
    19), (3, 36)))
data1: org.apache.spark.rdd.RDD[(Int, Int)] =
    ParallelCollectionRDD[37] at parallelize at <console>:16

scala> val data2 = sc.makeRDD(Array((1, "anil"), (2, "sunil")
    , (2, "kapil"), (4, "kgs")))
data2: org.apache.spark.rdd.RDD[(Int, String)] =
    ParallelCollectionRDD[38] at makeRDD at <console>:16

scala> data2.join(data1).collect()
14/09/05 14:10:25 INFO SparkContext: Starting job: collect at
    <console>:21
...
14/09/05 14:10:25 INFO SparkContext: Job finished: collect at
    <console>:21, took 0.065609053 s
res15: Array[(Int, (String, Int))] = Array((1,(anil,23)),
    (1,(anil,25)), (2,(sunil,19)), (2,(kapil,19)))
```

12. **cogroup**: The API usage is

```
RDD1.cogroup(RDD2, RDD3)
```

Returns an RDD of tuples with key k and list of values matching this key in 3 other RDDs.

Dataset

```
RDD1 = Array((1, 23), (2, 25), (3, 19), (4, 36))
RDD2 = Array((1, "anil"), (2, "sunil"), (3, "kapil"), (4, "
    kgs"))
RDD3 = Array((1, "CMU"), (1, "MIT"), (3, "GT"), (4, "UTA"))
```

Expected output

```
RDD = Array((4,(kgs, 36, UTA)), (1, (anil, 23, CMU, MIT)),
    (2, (sunil, 25, )), (3, (kapil, 19, GT)))
```

Spark Shell: Group all the RDDs based on the keys in each of them.

```
scala> val data1 = sc.parallelize(Array((1, 23), (2, 25), (3,
    19), (4, 36)))
data1: org.apache.spark.rdd.RDD[(Int, Int)] =
    ParallelCollectionRDD[62] at parallelize at <console>:16

scala> val data2 = sc.makeRDD(Array((1, "anil"), (2, "sunil"),
    (3, "kapil"), (4, "kgs")))
```

```
data2: org.apache.spark.rdd.RDD[(Int, String)] =
    ParallelCollectionRDD[63] at makeRDD at <console>:16

scala> val data3 = sc.parallelize(Array((1, "CMU"), (1, "MIT")
    , (3, "GT"), (4, "UTA")))
14/09/05 14:22:17 INFO ShuffleBlockManager: Deleted all files
    for shuffle 19
...
4/09/05 14:22:17 INFO ContextCleaner: Cleaned shuffle 17
data3: org.apache.spark.rdd.RDD[(Int, String)] =
    ParallelCollectionRDD[64] at parallelize at <console>:16

scala> data2.cogroup(data1,data3).collect()
14/09/05 14:22:21 INFO SparkContext: Starting job: collect at
    <console>:23
...
14/09/05 14:22:21 INFO SparkContext: Job finished: collect at
    <console>:23, took 0.075447687 s
res26: Array[(Int, (Iterable[String], Iterable[Int], Iterable
    [String]))] = Array((4,(ArrayBuffer(kgs),ArrayBuffer(36),
    ArrayBuffer(UTA))), (1,(ArrayBuffer(anil),ArrayBuffer(23),
    ArrayBuffer(CMU, MIT))), (2,(ArrayBuffer(sunil),
    ArrayBuffer(25),ArrayBuffer())), (3,(ArrayBuffer(kapil),
    ArrayBuffer(19),ArrayBuffer(GT))))
```

13. **cartesian**: The API usage is

```
RDD1.cartesian(RDD2)
```

Returns an RDD which is the Cartesian product of one RDD with the other. It contains the tuples of the form (a,b) where a ∈ RDD1 and b ∈ RDD2

Returns the Cartesian product of this RDD and another, that is, the RDD of all pairs of * elements (a, b) where a is in "this" and b is in "other".

Spark Shell: Find the Cartesian product of two arrays.

```
scala> val data1 = sc.parallelize(Array(1,2,3,4,5))
data1: org.apache.spark.rdd.RDD[Int] = ParallelCollectionRDD
    [1] at parallelize at <console>:12

scala> val data2 = sc.parallelize(Array('a','b'))
data2: org.apache.spark.rdd.RDD[Char] = ParallelCollectionRDD
    [2] at parallelize at <console>:12

scala> data1.cartesian(data2).collect()
14/09/05 14:30:54 INFO SparkContext: Starting job: collect at
    <console>:17
...
14/09/05 14:30:54 INFO SparkContext: Job finished: collect at
    <console>:17, took 0.238403285 s
res1: Array[(Int, Char)] = Array((1,a), (1,b), (2,a), (2,b),
    (3,a), (3,b), (4,a), (5,a), (4,b), (5,b))
```

14. **glom**: The API usage is

```
RDD.glom()
```

Returns an RDD by collecting all elements within each partition into an array.

Spark Shell: Create an RDD with elements of each partition as an array.

```
scala> val data = sc.makeRDD(Array(1,2,3,4,5,6,7,8),4)
data: org.apache.spark.rdd.RDD[Int] = ParallelCollectionRDD
    [4] at makeRDD at <console>:12

scala> data.glom().collect()
14/09/05 13:24:28 INFO SparkContext: Starting job: collect at
    <console>:15
...
14/09/05 13:24:28 INFO SparkContext: Job finished: collect at
    <console>:15, took 0.020780566 s
res3: Array[Array[Int]] = Array(Array(1, 2), Array(3, 4),
    Array(5, 6), Array(7, 8))
```

15. **intersection**: The API usage is

```
RDD1.intersection(RDD2)
```

Returns an RDD with elements equivalent to the intersection of two RDDs.

Spark Shell: Find the common elements between two RDDs

```
scala> val data1 = sc.makeRDD(Array(1,2,3,4,5,6))
data1: org.apache.spark.rdd.RDD[Int] = ParallelCollectionRDD
    [4] at makeRDD at <console>:12

scala> val data2 = sc.makeRDD(Array(5,6,7,8,4))
data2: org.apache.spark.rdd.RDD[Int] = ParallelCollectionRDD
    [5] at makeRDD at <console>:12

scala> data1.intersection(data2)
res1: org.apache.spark.rdd.RDD[Int] = MappedRDD[11] at
    intersection at <console>:17

scala> data1.intersection(data2).collect()
14/09/05 13:36:54 INFO SparkContext: Starting job: collect at
    <console>:17
...
14/09/05 13:36:54 INFO SparkContext: Job finished: collect at
    <console>:17, took 0.081608268 s
res2: Array[Int] = Array(4, 5, 6)
```

16. **union**: The API usage is

```
RDD1.union(RDD2)
```

Performs a Union operation on one RDD over another, returning an new RDD with all the elements of both the RDDs including the elements that are repeated.

Note: There is another operator ++ that performs the same function as the union method. The API usage is

```
RDD_1 ++ RDD_2
```

Spark Shell: Create a union of two RDDs

```
scala> val data1 = sc.makeRDD(Array(1,2,3,4,5))
data1: org.apache.spark.rdd.RDD[Int] = ParallelCollectionRDD
    [0] at makeRDD at <console>:12

scala> val data2 = sc.makeRDD(Array(6,7,8,9,10))
data2: org.apache.spark.rdd.RDD[Int] = ParallelCollectionRDD
    [1] at makeRDD at <console>:12

scala> data1.union(data2).collect()
14/09/05 13:13:54 INFO SparkContext: Starting job: collect at
    <console>:17
...
14/09/05 13:13:54 INFO SparkContext: Job finished: collect at
    <console>:17, took 0.183759737 s
res1: Array[Int] = Array(1, 2, 3, 4, 5, 6, 7, 8, 9, 10)

scala> (data1 ++ data2).collect()
14/09/05 13:16:36 INFO SparkContext: Starting job: collect at
    <console>:17
...
14/09/05 13:16:36 INFO TaskSchedulerImpl: Removed TaskSet
    1.0, whose tasks have all completed, from pool
res3: Array[Int] = Array(1, 2, 3, 4, 5, 6, 7, 8, 9, 10)
```

17. **reduce** : The API usage is

```
RDD.reduce(function)
```

Performs a reduce operation on the RDD by applying a commutative and associative binary operator.

Spark Shell: Add all the element of an RDD

```
scala> val data = sc.makeRDD(Array(1,2,3,4,5,6,7,8,9,10))
data: org.apache.spark.rdd.RDD[Int] = ParallelCollectionRDD
    [0] at makeRDD at <console>:12

scala> val newdata = data.reduce(_ + _)
14/09/05 12:36:49 INFO SparkContext: Starting job: reduce at
    <console>:14
...
14/09/05 12:36:49 INFO SparkContext: Job finished: reduce at
    <console>:14, took 0.174026012 s
```

```
newdata: Int = 55
```

18. **fold:** The API usage is

```
RDD.fold(zeroValue)(function)
```

Aggregates all the elements of an RDD by applying a function and zeroValue.

Spark Shell: Add all the contents of RDD using fold.

```
scala> val data = sc.makeRDD(Array(1,2,3,4))
data: org.apache.spark.rdd.RDD[Int] = ParallelCollectionRDD
    [1] at makeRDD at <console>:12

scala> val newdata = data.fold(0)(_ + _)
14/09/05 12:45:19 INFO SparkContext: Starting job: fold at
    <console>:14
...
14/09/05 12:45:19 INFO SparkContext: Job finished: fold at
    <console>:14, took 0.016982843 s
newdata: Int = 10
```

19. **collect:** The API usage is

```
RDD.collect()
```

Returns an array that contains all the RDD elements.

Spark Shell: list all the elements of an RDD

```
scala> val data = sc.makeRDD(Array("I", "Love", "Spark"))
data: org.apache.spark.rdd.RDD[String] =
    ParallelCollectionRDD[6] at makeRDD at <console>:12

scala> data.collect()
14/09/05 12:30:12 INFO SparkContext: Starting job: collect at
    <console>:15
...
14/09/05 12:30:12 INFO SparkContext: Job finished: collect at
    <console>:15, took 0.041148572 s
res1: Array[String] = Array(I, Love, Spark)
```

20. **count:** The API usage is

```
RDD.count()
```

Returns the number of elements in the RDD.

Spark Shell: Count the number of elements in the collection.

```
scala> val data = sc.makeRDD(Array(1,2,3,4,1,2,3,4,23,4))
data: org.apache.spark.rdd.RDD[Int] = ParallelCollectionRDD
    [2] at makeRDD at <console>:12
```

```
scala> data.count()
14/09/05 12:49:09 INFO SparkContext: Starting job: count at
     <console>:15
...
14/09/05 12:49:09 INFO SparkContext: Job finished: count at
     <console>:15, took 0.022548477 s
res0: Long = 10
```

21. **first**: The API usage is

    ```
    RDD.first()
    ```

    Returns the first element in the RDD.

    Spark Shell: Find the first element of a list

    ```
    scala> val data = sc.makeRDD(Array(1,2,3,4,5,6))
    data: org.apache.spark.rdd.RDD[Int] = ParallelCollectionRDD
        [0] at makeRDD at <console>:12

    scala> data.first()
    14/09/05 12:53:34 INFO SparkContext: Starting job: first at
        <console>:15
    ...
    14/09/05 12:53:34 INFO SparkContext: Job finished: first at
        <console>:15, took 0.02771088 s
    res0: Int = 1
    ```

22. **take**: The API usage is

    ```
    RDD.take(num)
    ```

    Returns an Array of first num elements in the RDD. Since RDD is partitioned, it
    checks the first partition and then looks for other partitions to satisfy the limit.

    Spark Shell: Select the first 3 elements of a collection.

    ```
    scala> val data = sc.makeRDD(Array(92, 90, 87, 85, 82, 80,
        79, 78, 60, 37))
    data: org.apache.spark.rdd.RDD[Int] = ParallelCollectionRDD
        [0] at makeRDD at <console>:12

    scala> data.take(3)
    14/09/05 12:58:48 INFO SparkContext: Starting job: take at
        <console>:15
    ...
    14/09/05 12:58:48 INFO SparkContext: Job finished: take at
        <console>:15, took 0.131502089 s
    res0: Array[Int] = Array(92, 90, 87)
    ```

23. **takeSample**: The API usage is

    ```
    RDD.takeSample(withReplacement, num)
    ```

Returns a Array of sample of the RDD elements. withReplacement is set to false if sampling is done without replacement else true, num indicates the size of the sample.

Spark Shell: Given a dataset of 10 elements, create a sample size of 5.

```
scala> val data = sc.makeRDD(Array(92, 90, 87, 85, 82, 80,
    79, 78, 60, 37))
data: org.apache.spark.rdd.RDD[Int] = ParallelCollectionRDD
    [5] at makeRDD at <console>:12

scala> data.takeSample(false, 5)
14/09/05 13:04:40 INFO SparkContext: Starting job: takeSample
    at <console>:15
...
14/09/05 13:04:40 INFO SparkContext: Job finished: takeSample
    at <console>:15, took 0.022600368 s
res5: Array[Int] = Array(37, 79, 60, 78, 80)
```

24. **takeOrdered**: The API usage is

```
RDD.takeOrdered(num)
```

Returns an ordered Array of an RDD sample of size num.

Spark Shell: Select 5 elements from a collection of 10 and sort them.

```
scala> val data = sc.makeRDD(Array(10,9,8,7,6,5,4,3,2,1))
data: org.apache.spark.rdd.RDD[Int] = ParallelCollectionRDD
    [0] at makeRDD at <console>:12

scala> data.takeOrdered(5)
14/09/05 13:09:13 INFO SparkContext: Starting job:
    takeOrdered at <console>:15
...
14/09/05 13:09:13 INFO SparkContext: Job finished:
    takeOrdered at <console>:15, took 0.234765781 s
res0: Array[Int] = Array(1, 2, 3, 4, 5)
```

25. **textFile**: The API usage is

```
sc.textFile(path, partitions)
```

Reads a file from an HDFS path, local file system, or any Hadoop supported file system URI and returns it as an RDD[String]

Spark Shell:

```
scala> sc.textFile("README.md")
14/09/05 09:54:35 INFO MemoryStore: ensureFreeSpace(32856)
    called with curMem=0, maxMem=309225062
14/09/05 09:54:35 INFO MemoryStore: Block broadcast_0 stored
    as values to memory (estimated size 32.1 KB, free 294.9 MB
    )
```

```
res0: org.apache.spark.rdd.RDD[String] = MappedRDD[1] at
    textFile at <console>:13
```

26. **saveAsTextFile**: The API usage is

```
RDD.saveAsTextFile(path)
```

Saves an RDD as text file using the string representations of its elements.

Note: saveAsTextFile in turn write the text content as a Hadoop file

```
HadoopFile[TextOutputFormal[NullWritable, Text]]
```

Spark Shell:

```
scala>val text = sc.textFile("README.md")
14/09/05 10:14:33 INFO MemoryStore: ensureFreeSpace(32856)
    called with curMem=32856, maxMem=309225062
14/09/05 10:14:33 INFO MemoryStore: Block broadcast_1 stored
    as values to memory (estimated size 32.1 KB, free 294.8 MB
    )
text: org.apache.spark.rdd.RDD[String] = MappedRDD[3] at
    textFile at <console>:12

scala>text.saveAsTextFile("savedREADME")
14/09/05 10:14:51 WARN NativeCodeLoader: Unable to load
    native-hadoop library for your platform... using builtin-
    java classes where applicable
...
14/09/05 10:14:51 INFO SparkContext: Job finished:
    saveAsTextFile at <console>:15, took 0.303714777 s
```

Saved file contents.

```
/savedREADME$  ls
part-00000  part-00001  _SUCCESS
```

27. **sequenceFile and saveAsSequenceFile**: The API usages are

```
sc.sequenceFile[K,V](path)
```

Returns an RDD of the Hadoop Sequence file with given key and value types.

```
RDD.saveAsSequenceFile(path)
```

Saves the RDD as a Hadoop Sequence file using the Writable types inferred by the RDD's key and value types. If the types are of Writable then their classes are used directly, else primitive data types like Int and Double are mapped to IntWritable and DoubleWritable, respectively. Similarly, Byte is mapped to BytesWritable and String to Text.

Spark Shell: Create an RDD and save it as SequenceFile. List the contents of the Sequence file.

```
scala> val data = sc.makeRDD(Array("Apache", "Apache"))
data: org.apache.spark.rdd.RDD[String] =
    ParallelCollectionRDD[7] at makeRDD at <console>:15

scala> val change = data.map(x => (x, "Spark"))
change: org.apache.spark.rdd.RDD[(String, String)] =
    MappedRDD[8] at map at <console>:17

scala> change.saveAsSequenceFile("changeout")
14/09/05 11:20:10 INFO SequenceFileRDDFunctions: Saving as
    sequence file of type (Text,Text)
14/09/05 11:20:10 INFO SparkContext: Starting job:
    saveAsSequenceFile at <console>:20

scala> val newchange = sc.sequenceFile[String, String]("
    changeout")
14/09/05 11:21:13 INFO MemoryStore: ensureFreeSpace(32880)
    called with curMem=98640, maxMem=309225062
14/09/05 11:21:13 INFO MemoryStore: Block broadcast_3 stored
    as values to memory (estimated size 32.1 KB, free 294.8 MB
    )
newchange: org.apache.spark.rdd.RDD[(String, String)] =
    MappedRDD[13] at sequenceFile at <console>:15

scala> newchange.collect()
14/09/05 11:21:17 INFO FileInputFormat: Total input paths to
    process : 4
14/09/05 11:21:17 INFO SparkContext: Starting job: collect at
    <console>:18
....
14/09/05 11:21:17 INFO SparkContext: Job finished: collect at
    <console>:18, took 0.02977156 s
res7: Array[(String, String)] = Array((Apache,Spark), (Apache
    ,Spark))
```

28. **objectFile and saveAsObjectFile:** The API usage are

```
sc.objectFile[K, V](path)
```

Loads an RDD that is saved as Sequence file containing serialized objects where K is NullWritable and V is BytesWritable

```
RDD.saveAsObjectFile(path)
```

Saves this RDD as a SequenceFile of serialized objects.

Spark Shell: Create a sequence of serialized objects and print the results

```
scala> val data = sc.makeRDD(Array(1,2,3,4))
data: org.apache.spark.rdd.RDD[Int] = ParallelCollectionRDD
    [0] at makeRDD at <console>:12
```

```
scala> val changeData = data.map(x => (x, "*" * x))
changeData: org.apache.spark.rdd.RDD[(Int, String)] =
    MappedRDD[1] at map at <console>:14

scala> val output = changeData.saveAsObjectFile("objectout")
14/09/05 11:35:50 INFO SequenceFileRDDFunctions: Saving as
    sequence file of type (NullWritable,BytesWritable)
14/09/05 11:35:50 INFO SparkContext: Starting job:
    saveAsObjectFile at <console>:16

....
14/09/05 11:35:51 INFO SparkContext: Job finished:
    saveAsObjectFile at <console>:16, took 0.319013396 s
output: Unit = ()

scala> val input = sc.objectFile[(Int, String)]("objectout")
14/09/05 11:36:28 INFO MemoryStore: ensureFreeSpace(32880)
    called with curMem=0, maxMem=309225062
14/09/05 11:36:28 INFO MemoryStore: Block broadcast_0 stored
    as values to memory (estimated size 32.1 KB, free 294.9 MB
    )
input: org.apache.spark.rdd.RDD[(Int, String)] =
    FlatMappedRDD[5] at objectFile at <console>:12

scala> input.collect()
14/09/05 11:36:35 INFO FileInputFormat: Total input paths to
    process : 4
14/09/05 11:36:35 INFO SparkContext: Starting job: collect at
    <console>:15
...
14/09/05 11:36:35 INFO SparkContext: Job finished: collect at
    <console>:15, took 0.052723211 s
res0: Array[(Int, String)] = Array((3,***), (4,****), (2,**),
    (1,*))
```

29. **countByKey**: The API usage is

```
RDD.countByKey()
```

Returns the count of the values for each key as (key, count) pair.

Spark Shell: Count the number of values of given (k,v) pairs

```
scala> val data = sc.makeRDD(Array(("a", 1),("b", 2),("a", 3)
    ))
data: org.apache.spark.rdd.RDD[(String, Int)] =
    ParallelCollectionRDD[0] at makeRDD at <console>:12

scala> data.countByKey()
14/09/05 11:48:37 INFO SparkContext: Starting job: countByKey
    at <console>:15
...
```

```
14/09/05 11:48:37 INFO SparkContext: Job finished: countByKey
    at <console>:15, took 0.217839736 s
res0: scala.collection.Map[String,Long] = Map(b -> 1, a -> 2)
```

30. **foreach**: The API usage is

```
RDD.foreach(function)
```

Applies a function to all the elements of RDD.

Spark Shell: Print all the elements of an RDD.

```
scala> val data = sc.makeRDD(Array(1,2,3,4)).foreach(x =>
    println(x))
14/09/05 11:56:03 INFO SparkContext: Starting job: foreach at
    <console>:12
...
3
2
1
4
...
14/09/05 11:56:03 INFO SparkContext: Job finished: foreach at
    <console>:12, took 0.181527602 s
```

References

1. Scala, "The Scala Programming Language," 2002. [Online]. Available: http://www.scala-lang.org/.
2. Twitter, Scalding, 2011. [Online]. Available: https://github.com/twitter/scalding.
3. Wensel, C. K. "Cascading: Defining and executing complex and fault tolerant data processing workflows on a hadoop cluster" (2008).
4. Cascading, "Cascading: Application Platform for Enterprise Big Data" [Online] Available: http://www.cascading.org/
5. Zaharia, Matei, et al. "Spark: cluster computing with working sets." Proceedings of the 2nd USENIX conference on Hot topics in cloud computing. 2010.
6. B. Hindman, A. Konwinski, M. Zaharia, and I. Stoica. A common substrate for cluster computing. In Workshop on Hot Topics in Cloud Computing (HotCloud) 2009, 2009.
7. Spark, Apache. [Online] Available: http://spark.incubator.apache.org/docs/latest/

Part II

Case studies using Hadoop, Scalding and Spark

Chapter 5

Case Study I: Data Clustering using Scalding and Spark

5.1 Introduction

Data mining is the process of discovering insightful, interesting, and novel patterns, as well as descriptive, understandable, and predictive models from large-scale data.

Explosive use of more data mining techniques began with the advent big data phenomenon and this explosion can be attributed to: First, to the size of the information which can scale up to a few Petabytes and Second, because the information tends to be more varied and extensive in its very nature and content. The complexity of the statistical techniques has also increased with large data sets. For example, if you have 40 or 50 million records of detailed customer information, discovering the customer location is not enough. The depth of analysis has increased to an extent, where we need to determine the demography of this customer base with insight into their interests to personalize the content as per the customer needs.

In the recent times, business has become more statistics driven and has reached a point where the analysis needs to be done in more real-time. For example, you might want to target offers/ads to the user currently online. Real-time analysis in the age big data you might want to form queries over huge sets of data that can scale-up to several Petabytes, and expect results in minutes. This is facilitated by the rise in distributed real-time computing technologies like Hadoop Impala, Apache Spark, etc.

Data Mining involves several knowledge discovery processes like *data extraction, data cleaning, data fusion, data reduction,* and *feature construction,* which is used for preprocessing. Postprocessing steps such as pattern and model interpretation, hypothesis confirmation, and generation and so on. Data Mining is also interdisciplinary in nature and uses the concepts from different areas such as database systems, statistics, machine learning, and pattern recognition. This knowledge discovery and data mining process is highly iterative and interactive and comprises of core

© Springer International Publishing Switzerland 2015
K.G. Srinivasa and A.K. Muppalla, *Guide to High Performance
Distributed Computing*, Computer Communications and Networks,
DOI 10.1007/978-3-319-13497-0_5

algorithms that enables one to gain fundamental insights and knowledge from massive data. The algebraic, geometric, and probabilistic viewpoints of data also play a key role in data mining.

5.2 Clustering

Clustering is process of grouping points into cluster according to a distance measure. The principle behind this is points that are closer to each other belong to the same cluster. Clustering is typically in advantage when the data size is very large. Clustering dataset is a collection of points that belong to some space. A space is just a universal set of points from which points in the dataset are drawn. Euclidean space has multiple properties that are useful for clustering. The Euclidean space points are real number vectors, length of vector is the number of dimensions of the space. The elements of the vector are called coordinates. All the spaces we can perform clustering have a distance measure, it is the distance between any two points in a space. The common euclidean distance measure is the square root of the sums of squares of distances between the coordinates of the points in each dimension. There are other distance measures for euclidean spaces like Manhattan distance (sum of the magnitudes of the differences in each dimension). Classical clustering applications involve low-dimensional euclidean spaces. Clustering algorithm will form clusters with small amount of data. Some applications involve euclidean spaces of very high dimensions, for example, it is challenging to cluster documents by their topic, based on the occurance of common, unusual words in the document. Another example, clustering movie goers by the type of movies they like. Distance measures between a pair of points:

1. Distance are non-negative and distance to itself is 0

2. Distance is symmetric; order of the points does not affect the distance

3. Distance measures obey the rule of triangle inequality; the distance from points x to y to z is never less than the distance going from x to z directly.

5.2.1 Clustering Techniques

Clustering techniques can be distinguished into two types of clustering techniques: Partition and Hierarchical.

- **Partition**: Given a database, the partition clustering algorithm partitions the data such that each cluster optimizes a clustering criterion. The disadvantage is the complexity of algorithms as some of them enumerate all possible groupings and

try to find a global optimum [15]. Even for a small dataset, the number of partitions is huge. Common solutions start with a random partition and is refined. A
better alternative is run the partition algorithm for different sets of initial points
and investigate whether all solutions lead to a final partition. Partition algorithm
locally improves a certain criterion. Initially, they compute the values of the similarity or distance and then pick the one that optimizes a criterion by ordering the
results. Most of them are considered as greedy-like algorithms.

- **Hierarchical**: Hierarchical algorithms create a hierarchical decomposition of the
  objects [16]. They are either agglomerative (bottom-up) or divisive (top-down):

  - *Agglomerative algorithms:* starts with each point being a cluster, using a dis-
    tance measure these individual cluster are merged. The clustering stops when
    all the points are in a single group or if the users chooses. These methods
    follow a greedy-like bottom-up merging.

  - *Divisive Algorithms:* starts with all the points belonging to one group and con-
    sequently splitting the group into smaller ones until each object falls in one
    cluster or when the user terminates. This method divides the objects into dis-
    joint groups at every step and this is done at every step until all the objects
    are divided in to separate clusters. This is similar to divide-and-conquer algo-
    rithms.

Apart from the two main categories of partitional and hierarchical clustering
algorithms, many other methods have emerged in cluster analysis, and are mainly
focused on specific problems or specific data sets available:

- *Density-Based Clustering:* . Density is defined as the number of objects in a
  particular neighborhood of data objects. Density-based clustering method
  groups objects according the density objective functions. The cluster contin-
  ues to grow as long as the number of objects in that neighborhood exceeds
  some parameter. This is different from partition algorithms that use iterative
  relocation of points.

- *Grid-Based Clustering:* Data that represent the geometric structure, relationships, properties, and operations between objects is called spatial data. The
  goal of these algorithms is to quantize the data into a number of cells and
  work on the objects in these cells. They do not relocate the points but build
  hierarchical levels of groups of objects. These algorithms are similar to hierarchical algorithms but the merging of groups is defined by a parameter and
  not a distance measure.

- *Model-Based Clustering:* They can be either partition-based or hierarchical
  depending on the structure or model they hypothesize about the dataset and
  the way they refine the model to identify partitions. They find good approximations that best fits the data. They are similar to density-based models in
  that they grow the model so that it is better than the previous model iteration.

However, there are scenarios where they start with fixed number of cluster and these are not usually density based.

– *Categorical Data Clustering:* These algorithms are useful in clustering where any form of distance measure cannot be applied. The principles are close to partition-based and hierarchical methods.

Each category can be further divided in subcategories, example, density-based clustering with respect to geographical data. An exception is categorical data approaches. Data Visualization is not straight forward, in the sense there is no obvious structure to the data, hence approaches mainly involve concepts like co-occurrences in tuples. There are datasets with a mixture of attribute types such as the United States census dataset, datasets used in schema discovery. The common clustering methods focus of datasets where all the attributes are of a single type.

What makes a clustering algorithm effective and efficient? The answer is not clear. They are data dependent while one method works well on one type of data but fails miserably on another because the algorithms depend on the size, dimensionality, and the objective function of the data. The characteristics of a good clustering technique are:

· *Scalability:* Algorithm needs to perform well with large number of objects.

· *Analyze multiple attribute types:* Algorithm needs to accommodate single and multiple attribute types.

· *Find arbitrary-shaped clusters:* The shapes usually represent the kinds of clusters an algorithm can derive, this is important in choosing a model. Different algorithms will be biased on different characteristics or shapes and it is not easy to determine the shape or the corresponding bias. When there are categorical attributes, it is difficult to determine cluster structures.

· *Minimum requirements for input parameters:* Algorithms require some user-defined parameters such as the number of clusters, in order to analyze the data. However with large datasets and higher dimensions, it is advisable to provide only limited guidance in order to avoid bias over the result.

· *Handling of noise:* Algorithm should be able to account for deviations. Deviations are normally called outliers as they depart from the accepted behavior of the objects. Deviation detection is normally handled separately.

· *Sensitivity to the order of input records:* The algorithm is expected to agnostic toward the order of input. Certain algorithms behave differently when input is presented in different order. The order mostly affects those algorithms that require single scan over the dataset, leading to locally optimal solutions.

· *High dimensionality of data:* Many algorithms cannot handle high-dimensional data. It is a challenge to cluster high-dimensional datasets, such as the U.S. census data set which contains more attributes. The appearance of large number of attributes is often termed as the curse of dimensionality due to:

1. As the number of attributes becomes larger, the amount of resources required to store or represent them grows.

2. The distance of a given point from the nearest and furthest neighbor is almost the same, for a wide variety of distributions and distance functions.

These two properties affect the efficiency of a clustering algorithm since it require more time to proces the data while the resulting cluster are of poor quality.

5.2.2 Clustering Process

Data Collection: Involves extraction of relevant data objects from different data sources. These data objects are distinguished by their values.

Initial Screening: Also known as a data cleaning process, commonly defined in Data Warehousing.

Representation: In most cases, the dataset cannot be used as is needs to transformed to abide by the clustering algorithm. For example, a similarity measure is used to examine the characteristics and dimensionality of data.

Clustering Tendency: Smaller datasets are checked to see if the data can be clustered. This stage is often ignored, especially in large datasets.

Clustering Strategy: Involves choosing the right clustering algorithm and initial parameters.

Validation: This is a very important stage which is often based on manual examination and visual techniques. However, as the size of the data grows the number of ways to compare with preconceived ideas are reduced.

Interpretation: This stage includes combining clustering and other forms of analysis like classification to draw conclusions.

5.2.3 K-Means Algorithm

The best known family of clustering algorithms is the *K-Means* algorithms. These algorithms assume the points are distributed in an Euclidean space and the number of clusters k is known in advance.

Given a multidimensional space $D = \{x_i\}_{i=1}^n$, with n points and k desired clusters. The goal of the clustering algorithm is to partition the dataset into k groups or clusters denoted by $C = \{C_1, C_2, C_3, ...C_n\}$. Further, for each cluster C_i there exists a point called the centroid that summarizes the cluster, denoted by μ_i of all the points in the cluster.

$$\mu_i = \frac{1}{n} \sum_{x_j \in C_i} x_j$$

where $n_i = |C_i|$ is the number of points in cluster C_i.

A brute-force approach to good clustering is finding all the possible partitions on n points into k clusters, evaluate some optimization between them, retain those k clusters with the best score. K-Means algorithm initializes the cluster centroid by randomly generating k points in the data space n. Each iteration in K-Means consists of two steps:

1. Cluster Assignment

2. Centroid Update.

After choosing k cluster centroids, each point $x_j \in D$ is assigned to the closest centroid, which results in a cluster where each cluster C_i consists of points closer to the centroid μ_i than any other cluster centroid. The following equation represents the assignment: Each point x_j is assigned to cluster C_p.

$$p = argmin_{i=1}^k \left\{ ||x_j - \mu_i||^2 \right\}$$

In the Centroid Update step, given a set of clusters C_i, i = { 1, ..., k} a new centroid/mean is calculated for each cluster from the points in C_i. The Cluster Assignment and Centroid Update steps are iteratively executed until we find a local minima. Practically, one can assume the K-Means algorithm has converged if the centroids do not change from one iteration to the next.

The pseudocode for the K-Means algorithm is given in 1. In terms of computational complexity, K-Means cluster assignment step takes O(nkd) time, because we have to find the distance from each point to each of the k cluster centroids, which takes d operations in d dimensions. The Centroid Update step takes O(nd) time because we have to update all of n d dimensional points. Assuming there are t iterations, the total time for K-Means is given as $O(tnkd)$.

Algorithm 1: K-Means

K-Means(D, k, $\in$)

$t = 0$

Ramdomly intitialize k centroids $\mu_1^t, \mu_2^t, ..., \mu_k^t$

repeat

    $t \longleftarrow t + 1$

    $C_j \longleftarrow \phi$ for all j = 1,...,k

    // Centroid Assign

    foreach $x_{j|\in D}$ **do**

        $p \longleftarrow argmin_i \left\{ ||x_j - \mu_i^t||^2 \right\}$

        $C_p \longleftarrow C_p \cup x_j$

    // Centroid Update

    foreach $i = 1$ *to* k **do**

        $\mu_i^t \longleftarrow \frac{1}{|C_j|} \sum_{x_j \in C_i} x_j$

until $\sum_{i=1}^{k} ||\mu_i^t - \mu_i^{t-1}||^2$;

5.2.4 Simple K-Means Example

Consider the following data set consisting of the scores of two variables on each of seven individuals as shown in Table 5.1.

Table 5.1: Marks of 7 students for subjects A, B

Subject	A	B
1	1.0	1.0
2	1.5	2.0
3	3.0	4.0
4	5.0	7.0
5	3.5	5.0
6	4.5	5.0
7	3.5	4.5

This data set is to be grouped into two clusters. As a first step in finding a sensible initial partition, let the A & B values of the two individuals furthest apart (using the Euclidean distance measure), define the initial cluster means, giving Table 5.2.

Table 5.2: First Centroids

	Individual	Mean Vector (centroid)
Group 1	1	(1.0, 1.0)
Group 2	4	(5.0, 7.0)

The remaining individuals are now examined in sequence and allocated to the cluster to which they are closest, in terms of Euclidean distance to the cluster mean. The mean vector is recalculated each time a new member is added. This leads to the following series of steps as shown in Table 5.3.

Table 5.3: Iteration

	Cluster 1		Cluster 2	
Step	Individual	Mean Vector (centroid)	Individual	Mean Vector (centroid)
1	1	(1.0, 1.0)	4	(5.0, 7.0)
2	1, 2	(1.2, 1.5)	4	(5.0, 7.0)
3	1, 2, 3	(1.8, 2.3)	4	(5.0, 7.0)
4	1, 2, 3	(1.8, 2.3)	4, 5	(4.2, 6.0)
5	1, 2, 3	(1.8, 2.3)	4, 5, 6	(4.3, 5.7)
6	1, 2, 3	(1.8, 2.3)	4, 5, 6, 7	(4.1, 5.4)

Now the initial partition has changed, and the two clusters at this stage have the following characteristics as shown in Table 5.4.

Table 5.4: Changed Clusters

	Individual	Mean Vector (centroid)
Cluster 1	1, 2, 3	(1.8, 2.3)
Cluster 2	4, 5, 6, 7	(4.1, 5.4)

But we cannot yet be sure that each individual has been assigned to the right cluster. So, we compare each individuals distance to its own cluster mean and to that of the opposite cluster. And we find in Table 5.5 only individual 3 is nearer to the mean of the opposite cluster (Cluster 2) than its own (Cluster 1). In other words, each individual's distance to its own cluster mean should be smaller that the distance to the other cluster's mean (which is not the case with individual 3). Thus, individual 3 is relocated to Cluster 2 resulting in the new partition as shown in Table 5.6.

Table 5.5: Distance to Clusters

Individual	Distance to mean (centroid) of Cluster 1	Distance to mean (centroid) of Cluster 2
1	1.5	5.4
2	0.4	4.3
3	2.1	1.8
4	5.7	1.8
5	3.2	0.7
6	3.8	0.6
7	2.8	1.1

Table 5.6: Final Clusters

	Individual	Mean Vector Centroid
Cluster 1	1, 2	(1.3, 1.5)
Cluster 2	3, 4, 5, 6, 7	(3.9, 5.1)

The iterative relocation would now continue from this new partition until no more relocations occur. However, in this example each individual is now nearer its own cluster mean than that of the other cluster and the iteration stops, choosing the latest partitioning as the final cluster solution.

Also, it is possible that the K-Means algorithm would not find a final solution. In this case, it would be a good idea to consider stopping the algorithm after a prechosen maximum of iterations.

5.3 Implementation

The objective here is to implement K-Means clustering algorithm using Scalding and Spark frameworks to find clusters in the given revenue data of several Organizations.

Dataset The dataset contains information about 30 organizations. Each row represents an organization and columns represent the size of the organization and revenue generated in millions.

- Dimension 0: Size of the organization
- Dimension 1: Revenue of the organization in millions

A sample data is given in Table 7.11

Table 5.7: Dataset

Id	Size	Revenue
0	6882	3758
1	9916	5478
2	9368	3521
3	6386	4809
4	7930	4260
5	8997	5141
6	9401	3874
7	6579	3804
8	9454	3839
9	8317	5929
10	2683	4757

This dataset when plotted looks as in Figure. 5.1

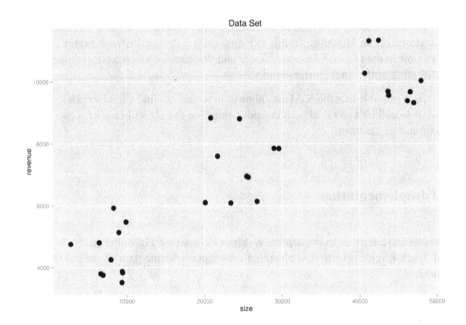

Fig. 5.1: Size and Revenue Data Distribution

5.3.1 Scalding Implementation

The data clustering, using the Apache Mahout K-Means library, is implemented by the following steps:

- Data Preprocessing
- Determining the initial centroids
- Applying K-Means Algorithm to get the final centroids
- Post Processing

Data Pre-processing

Step 1: Initialize Scalding Job, here `InputJob` extends `Job` class that implements the necessary I/O formatting and all other hadoop configuration. Initialize the input from a Tab Separated Values file using Fields Api of scalding

```
import com.twitter.scalding._
class InputJob(args: Args) extends Job(args) {
val userFeatures = Tsv("input.tsv", ('id, 'x, 'y))
}
```

Step 2: Read the data

```
import com.twitter.scalding._
class InputJob(args: Args) extends Job(args) {
val userFeatures = Tsv("input.tsv", ('id, 'x, 'y))
    .read
}
```

Step 3: Vectorizing Data — `mapTo` function is applied to vectorize the data in the map task.

The definition of the `mapTo` function is as follows:

```
pipe.mapTo(existingFields -> additionalFields){function}
```

`mapTo` is equivalent to mapping and then projecting to the new fields, but is more efficient. Thus, the following two lines produce the same result:

```
pipe.mapTo(existingFields -> additionalFields){ ... }
pipe.map(existingFields -> additionalFields){ ... }
    .project(additionalFields)
```

Following code performs the vectorization of the data that is required for processing using the K-Means Mahout algorithm. Given the dataset has the columns [id, size, revenue], the data has been read and stored as [id, x, y] fields, respectively. We need to perform a mapping task such that the dataset is vectorized by using the id and creating vector of x and y. The resulting data after the map transformation we get `user` (the company id's) and `vector` fields.

Initially, all the data that is read into the variable `userFeatures` is of type `java.lang.String`. The `mapTo` function replaces the fields [id, x, y] to [user, vector]. Since Scala is a fucntional programming language, everything is treated as a function of something. We write a function `f` such that it takes the parameters (id, x, y) maps each input row and converts the fields x and y into a vector using the `DenseVector` class. The resulting vector is then converted to a named vector with the corresponding id of each row.

```scala
import org.apache.mahout.math.{NamedVector, DenseVector,
            VectorWritable}
import com.twitter.scalding._
class InputJob(args: Args) extends Job(args) {
val userFeatures = Tsv("input.tsv", ('id, 'x, 'y))
  .read
  .mapTo(('id, 'x, 'y) -> ('user, 'vector)){
    f : (String, String, String) =>
    val user = f._1
    val vector = Array(f._2.toDouble, f._3.toDouble)
    val namedVector = new NamedVector(
            new DenseVector(vector), user)
    val vectorWritable = new VectorWritable(namedVector)
    (new Text(user), vectorWritable)
  }
}
```

DenseVector: Implements vector from an array of doubles

```scala
DenseVector(double[] values)
```

From the constructor definition of the `DenseVector` class, we understand that the features to be vectorized need to be an Array. We use the Scala builtin `Array` library to add the fieds x and y.

```scala
val vector = Array(f._2.toDouble, f._3.toDouble)
```

We use `Named Vector` class to convert the vector and associate a name as key, whose constructor definition is:

```scala
NamedVector(Vector delegate, String name)
```

```scala
val namedVector = new NamedVector(
            new DenseVector(vector), user)
```

We then move to converting the `namedVector` to an instance of `VectorWritable`, which is the required format for Mahout K-Means algorithm.

```
val vectorWritable = new VectorWritable(namedVector)
```

Step 5: Write the vectorized data to a Hadoop SequenceFile.

```
WritableSequenceFile(p: String, f: Fields,
    sinkMode: SinkMode = cascading.tap.SinkMode.REPLACE)
```

Here p = Path, f = fields, usually a tuple to indicate the [Key, Value].

```
import org.apache.mahout.math.{NamedVector, DenseVector,
                VectorWritable}
import com.twitter.scalding._
class InputJob(args: Args) extends Job(args) {
val userFeatures = Tsv("input.tsv", ('id, 'x, 'y))
  .read
  .mapTo(('id, 'x, 'y) -> ('user, 'vector)){
    f : (String, String, String) =>
    val user = f._1
    val vector = Array(f._2.toDouble, f._3.toDouble)
    val namedVector = new NamedVector(
                new DenseVector(vector), user)
    val vectorWritable = new VectorWritable(namedVector)
    (new Text(user), vectorWritable)
  }
}
val out = WritableSequenceFile[Text, VectorWritable](
    mahout_vectors, ('user , 'vector))
userFeatures.write(out)
```

Sample Output The vectorized data is shown in Table 5.8

Table 5.8: vectorized dataset

user	vector
0	0:0:6882.0,1:3758.0
1	1:0:9916.0,1:5478.0
2	2:0:9368.0,1:3521.0
3	3:0:6386.0,1:4809.0
4	4:0:7930.0,1:4260.0
5	5:0:8997.0,1:5141.0
6	6:0:9401.0,1:3874.0
7	7:0:6579.0,1:3804.0
8	8:0:9454.0,1:3839.0
9	9:0:8317.0,1:5929.0
10	10:0:2683.0,1:4757.0

Determining the Initial Centroids.

Apache Mahout implements a randomized initial cluster selection by using the Reservoir Sampling method: Reservoir sampling is a family of randomized algorithms for randomly choosing a sample of k items from a list S containing n items, where n is either a very large or unknown number.

Step 6: The class used is called `RandomSeedGenerator`. Given an Input Path containing a `org.apache.hadoop.io.SequenceFile`, randomly select k vectors and write them to the output file as `org.apache.mahout. clustering. kmeans.Kluster` representing the initial centroid to use. The function we focus is called `buildRandom`. The function definition is:

```
buildRandom(Configuration conf, Path input,
    Path output, int k, DistanceMeasure measure)
```

- **conf**: hadoop configuration settings, serialization is the process of translating data structures or object state into a format that can be stored (for example, in a file or memory buffer, or transmitted across a network connection link) and reconstructed later in the same or another computer environment.

  ```
  val conf = new Configuration
  conf.set("io.serializations",
      "org.apache.hadoop.io.serializer.
                          JavaSerialization," +
      "org.apache.hadoop.io.serializer.
                          WritableSerialization")
  ```

- **input**: Path, which is of type `org.apache.hadoop.fs.Path`, this points to the Hadoop Sequence file that was contains the vectorized data.

  ```
  val vectorsPath = new Path("data/kmeans/mahout_vectors")
  ```

- **output**: Path, which is of type `org.apache.hadoop.fs.Path`, this points to the location where the initial set of centroids are written to. The file is of the type `org.apache.mahout.clustering.kmeans.Kluster.` that contains the `center, clusterId, distance measure`. The third field explains that the cluster centers are determined using the effective distance between the points. This distance can be measured in several ways which we will see in the next step.

  ```
  val inputClustersPath = new Path("data/kmeans/random_centroids")
  ```

- **k**: K, number of clusters which are the required input in the K-Means Algorithms (K in K-Means is the number clusters!)

- **distanceMeasure**: We find the cluster centers as a function of EuclideanDistanceMeasure. the Euclidean distance or Euclidean metric is the "ordinary" distance between two points that one would measure with a ruler, and is given by the Pythagorean formula. By using this formula as distance.

$$d(p,q) = \sqrt{(q_1 - p_1)^2 + (q_2 - p_2)^2 ... (q_n - p_n)^2} = \sqrt{\sum_{i=1}^{n} (q_i - p_i)^2}$$

Mahout provides a readymade class for performing EuclideanDistanceMeasures.

```
import org.apache.mahout.common.distance.EuclideanDistanceMeasure
val distanceMeasure = new EuclideanDistanceMeasure
```

The final routine looks like this:

```
import org.apache.mahout.clustering.kmeans.RandomSeedGenerator
import org.apache.mahout.common.distance.EuclideanDistanceMeasure
import org.apache.hadoop.conf.Configuration
import org.apache.hadoop.fs.Path

val conf = new Configuration
conf.set("io.serializations",
    "org.apache.hadoop.io.serializer.JavaSerialization, " +
    "org.apache.hadoop.io.serializer.WritableSerialization")
val inputClustersPath = new Path("data/kmeans/random_centroids")
val distanceMeasure = new EuclideanDistanceMeasure
val vectorsPath = new Path("data/kmeans/mahout_vectors")

RandomSeedGenerator.buildRandom(conf, vectorsPath,
    inputClustersPath, 3, distanceMeasure)
```

Sample output of the selected centroids, bear in mind that the output is a Sequential file and cannot be read by any text editor; therefore, you need to convert the File format to a more readable one. Scalding provides very simple one line solution to doing this. Write a simple Scalding Job as follows

- **Read the Sequence File**: The data is read from the path "data/kmeans/random_centroids". The Data is stored as two columns; [ClusterId, Cluster], these are read and stored in Scalding Fields type, clusterId, which is of the type org.apache.hadoop.io.Text and cluster of type org.apache.mahout.clustering.iterator.ClusterWritable.

```
import org.apache.mahout.clustering.iterator.ClusterWritable
import org.apache.hadoop.io.Text
import com.twitter.scalding._

class RandomCentroidJob(args: Args) extends Job(args) {
    val randomCentroids = WritableSequenceFile[Text, ClusterWritable](
    "data/kmeans/random_centroids", ('clusterId, 'cluster))
    .read
```

- **Convert the fields form hadoop file formats to readable**. Any form of `Fields` conversion we use the `mapTo` function. It takes the [`clusterId, cluster`] of each row. Since `cluster` if of type `ClusterWritable`, we need to use the built in class function to retrieve the center. This is done by

```
val center = x._2.getValue.getCenter
```

The converted fields are then tupled as (`cid, center`).

```
.mapTo(('clusterId, 'cluster) -> ('cid, 'center)) {
    x: (Text, ClusterWritable) =>
    val cid = x._1.toString.toDouble
    val center = x._2.getValue.getCenter
    println("cluster center -> " + center)
    (cid, center)
    }
    .write(Tsv("random_centroids.tsv"))
```

The final Code looks like the following:

```
import org.apache.mahout.clustering.iterator.ClusterWritable
import org.apache.hadoop.io.Text
import com.twitter.scalding._

class RandomCentroidJob(args: Args) extends Job(args) {
    val randomCentroids = WritableSequenceFile[Text, ClusterWritable](
    "data/kmeans/random_centroids", ('clusterId, 'cluster))
    .read
    .mapTo(('clusterId, 'cluster) -> ('cid, 'center)) {
    x: (Text, ClusterWritable) =>
    val cid = x._1.toString.toDouble
    val center = x._2.getValue.getCenter
    println("cluster center -> " + center)
    (cid, center)
    }
    .write(Tsv("random_centroids.tsv"))
}
```

Output

ClusterId	Cluster in Vector form
18	18:{0:48017.0,1:10037.0}
25	25:{0:23403.0,1:6090.0}
5	5:{0:8997.0,1:5141.0}

Random centers selected can be represented as in Figure 5.2

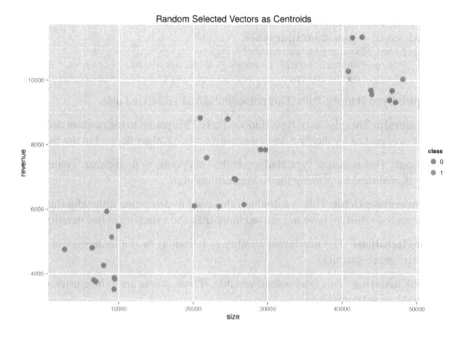

Fig. 5.2: Random Initial Centroids selected as class 1

Applying the K-Means Algorithm

Step 7: The K-Means clustering algorithm may be run using a command-line invocation on `KMeansDriver.main` or by making a Java call to `KMeansDriver.run()`.

The function defintion looks like this:

```java
public static void run(
    Configuration conf,
    Path input,
    Path clustersIn,
    Path output,
    double convergenceDelta,
    int maxIterations,
    boolean runClustering,
    double clusterClassificationThreshold,
    boolean runSequential)
```

- **conf**: hadoop configuration settings, serialization is the process of translating data structures or object state into a format that can be stored (for example, in a file or memory buffer, or transmitted across a network connection link) and reconstructed later in the same or another computer environment.

```
val conf = new Configuration
conf.set("io.serializations",
    "org.apache.hadoop.io.serializer.JavaSerialization," +
    "org.apache.hadoop.io.serializer.WritableSerialization")
```

- **input**: Type Hadoop Path. This points the initial vectorized data.

- **clustersIn**: This also is of type Hadoop Path. This points to the random centroids that was picked by the `RandomSeedGenerator` class defined in the Step 6.

- **output**: This is also of type Hadoop Path. This points to final cluster centers that are determined by running the K-Means algorithm.

- **convergenceDelta**: This is a double value used to determine if the algorithm has converged (clusters have not moved more than the value in the last iteration)

- **maxIterations**: The maximum number of iterations to run, independent of the convergence specified.

- **runClustering**: This is a boolean variable, if true, points are to be clustered after iterations are completed

- **clusterClassificationThreshold**, Is a clustering strictness / outlier removal parameter. Its value should be between 0 and 1. Vectors having pdf below this value will not be clustered.

- **runSequential**: This is a boolaean variable. If true, a sequential program is used to find the final centroids.

The final code looks like this

```
KMeansDriver.run(
    conf,
    new Path("data/kmeans/mahout_vectors"),// INPUT
    new Path("data/kmeans/random_centroids"),// initial centroids
    new Path("data/kmeans/result_cluster"),// OUTPUT_PATH
    0.01,                   //convergence delta
    20,                     // MAX_ITERATIONS
    true,                   // run clustering
    0,                      // cluster classification threshold
    false)                  // run Sequential
```

Post Processing

Step 8: After the execution of the K-Means Algorithm, the Mahout implementation writes the final cluster centroids to a file that has the suffix `-final`.

The code used to determine the initial clusters is used with path to read file changed.

First, we construct the path to the final clusters

```
val finalClusterPath = "data/kmeans/result_cluster" + "/*-final"
```

Final code looks like below:

```
import org.apache.mahout.clustering.iterator.ClusterWritable
import org.apache.hadoop.io.Text
import com.twitter.scalding._

class FinalCentroidsJob(args: Args) extends Job(args) {
    val finalClusterPath = "data/kmeans/result_cluster" +
                                   "/*-final"
    val finalCluster = WritableSequenceFile[
                                   Text,
                                   ClusterWritable](
                                   finalClusterPath,
                                   ('clusterId, 'cluster))
    .read
    .mapTo(('clusterId, 'cluster) -> ('cid, 'center)) {
    x: (Text, ClusterWritable) =>
    val cid = x._1.toString.toDouble
    val center = x._2.getValue.getCenter
    println("cluster center -> " + center)
    (cid, center)
    }
    .write(Tsv("final_centroids.tsv"))
}
```

The final centroids are:

ClusterId	Center(x)	Center(y)
0	7810.27	4470.0
1	24720.3	7313.9
2	44470.9	10068.7

These points are represented as squares in Figure 5.3

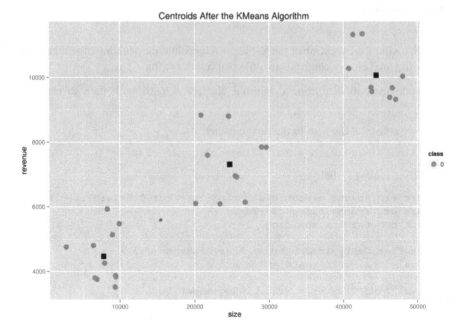

Fig. 5.3: Centroids After the K-Means Algorithm

Step 9: Now that we have the final centroids, we need find the distances for each vector against each of the centroids. Find the minimum distance of the set of distances for each input vector. The centroid associated with the minimum distance is the correct centroid.

The following iterative algorithm looks like this:

```
for input in range [0 to end of inputs]
    minDist = \inf
    for centroid in rage[0 to end of centroids]
        dist = EuclideanDistanceMeasure(input, centroid)
        if dist < minDist
            minDist = dist
    print input, centroid, minDist
```

This algorithm forms a perfect scenario for Map and Reduce kind of operations. We break the above alogrithm build using scalding as follows:

It is easier to distribute the job on the cluster if the data required for the computation is available for each input row, that way each row of input can be worked in the Map and produce a result which can be collated later to make sense in the Reduce task.

- Create a cross product of the input dataset and final centroid set, that is if there are N input rows and M centroid rows, then the total number of rows become NxM. Scalding provides a simple function for this:

```
pipe1.crossWithTiny(pipe2)
```

Performs the cross product of two pipes. The right (pipe2) is replicated to all the nodes, and the left is not moved at all. Therefore, the "tiny" part should be on the right.existingFields

First we read the input dataset and cross this data with the finalCluster computed in the previous step. The sample output is show in figure 5.9

```
import com.twitter.scalding._
class InputJob(args: Args) extends Job(args) {
val userVectors  = WritableSequenceFile[Text, VectorWritable]
               (mahout_vectors, ('user , 'vector))
    .crossWithTiny(finalCluster)
    .write(Tsv("crosswithtiny.tsv"))
```

Table 5.9: Sample out of CrossWithTiny

id	input vectors	ClusterId	ClusterCenters
0	0:{0:6882.0,1:3758.0}	0.0	{0:7810.2727,1:4470.0}
0	0:{0:6882.0,1:3758.0}	1.0	{0:24720.3,1:7313.9}
0	0:{0:6882.0,1:3758.0}	2.0	{0:44470.8889,1:10068.77}
1	1:{0:9916.0,1:5478.0}	0.0	{0:7810.2727,1:4470.0}
1	1:{0:9916.0,1:5478.0}	1.0	{0:24720.3,1:7313.9}
1	1:{0:9916.0,1:5478.0}	2.0	{0:44470.8889,1:10068.77}
2	2:{0:9368.0,1:3521.0}	0.0	{0:7810.2727,1:4470.0}
2	2:{0:9368.0,1:3521.0}	1.0	{0:24720.3,1:7313.9}
2	2:{0:9368.0,1:3521.0}	2.0	{0:44470.8889,1:10068.77}
3	3:{0:6386.0,1:4809.0}	0.0	{0:7810.2727,1:4470.0}
3	3:{0:6386.0,1:4809.0}	1.0	{0:24720.3,1:7313.9}

- Perform a map operation to determine the distances for a vector and a centroid for each row of the above data. The sample output is shown in Table 5.10

```
import org.apache.mahout.common.distance.EuclideanDistanceMeasure
import org.apache.mahout.math.{NamedVector, DenseVector, VectorWritable}
import org.apache.mahout.clustering.iterator.ClusterWritable
import com.twitter.scalding._

class InputJob(args: Args) extends Job(args) {
val userVectors  = WritableSequenceFile[Text, VectorWritable]
               (mahout_vectors, ('user , 'vector))
    .crossWithTiny(finalCluster)
    .map(('center, 'vector) -> 'distance) {
        x:(DenseVector, VectorWritable) =>
```

```
        (new EuclideanDistanceMeasure()).distance(x._1,
                                                   x._2.get)
    }
    .write(Tsv("Distance.tsv"))
```

Table 5.10: Compute EuclideanDistance(vector, centroid)

id	Input vectors	ClusterId	ClusterCenters	Distance
0	0:{0:6882.0,1:3758.0}	0.0	{0:7810.2727,1:4470.0}	1169.88696
0	0:{0:6882.0,1:3758.0}	1.0	{0:24720.3,1:7313.9}	18189.26546
0	0:{0:6882.0,1:3758.0}	2.0	{0:44470.8889,1:10068.77}	38114.96449
1	1:{0:9916.0,1:5478.0}	0.0	{0:7810.2727,1:4470.0}	2334.55917
1	1:{0:9916.0,1:5478.0}	1.0	{0:24720.3,1:7313.9}	14917.701846
1	1:{0:9916.0,1:5478.0}	2.0	{0:44470.8889,1:10068.77}	34858.508086
2	2:{0:9368.0,1:3521.0}	0.0	{0:7810.2727,1:4470.0}	1824.038117
2	2:{0:9368.0,1:3521.0}	1.0	{0:24720.3,1:7313.9}	15813.892849
2	2:{0:9368.0,1:3521.0}	2.0	{0:44470.8889,1:10068.77}	35708.34082
3	3:{0:6386.0,1:4809.0}	0.0	{0:7810.2727,1:4470.0}	1464.069013
3	3:{0:6386.0,1:4809.0}	1.0	{0:24720.3,1:7313.9}	18504.62308

- Convert the user, that is the row user to double so that it helps the consequent grouping function and the vector to String printable format as shown in Table 5.11

```
import org.apache.mahout.common.distance.EuclideanDistanceMeasure
import org.apache.mahout.math.{NamedVector, DenseVector, VectorWritable}
import org.apache.mahout.clustering.iterator.ClusterWritable
import cascading.pipe.joiner._
import com.twitter.scalding._
class InputJob(args: Args) extends Job(args) {
val userVectors  = WritableSequenceFile[Text,VectorWritable]
                    (mahout_vectors, ('user , 'vector))
    .crossWithTiny(clusterCenter)
    .map(('center, 'vector) -> 'distance) {
        x:(DenseVector, VectorWritable) =>
            (new EuclideanDistanceMeasure()).distance(x._1,
                                                      x._2.get)
    }
    .map(('user, 'vector) -> ('user, 'vector)){
        x : (Text, VectorWritable) =>
                (x._1.toString.toDouble, x._2.toString)
    }
    .write(Tsv("Mapped.tsv"))
```

Table 5.11: Compute EuclideanDistance(vector, centroid)

id	Input vectors	ClusterId	ClusterCenters	Distance
0.0	0:{0:6882.0,1:3758.0}	0.0	{0:7810.2727,1:4470.0}	1169.88696
0.0	0:{0:6882.0,1:3758.0}	1.0	{0:24720.3,1:7313.9}	18189.26546
0.0	0:{0:6882.0,1:3758.0}	2.0	{0:44470.8889,1:10068.77}	38114.96449
1.0	1:{0:9916.0,1:5478.0}	0.0	{0:7810.2727,1:4470.0}	2334.55917
1.0	1:{0:9916.0,1:5478.0}	1.0	{0:24720.3,1:7313.9}	14917.701846
1.0	1:{0:9916.0,1:5478.0}	2.0	{0:44470.8889,1:10068.77}	34858.508086
2.0	2:{0:9368.0,1:3521.0}	0.0	{0:7810.2727,1:4470.0}	1824.038117
2.0	2:{0:9368.0,1:3521.0}	1.0	{0:24720.3,1:7313.9}	15813.892849
2.0	2:{0:9368.0,1:3521.0}	2.0	{0:44470.8889,1:10068.77}	35708.34082
3.0	3:{0:6386.0,1:4809.0}	0.0	{0:7810.2727,1:4470.0}	1464.069013
3.0	3:{0:6386.0,1:4809.0}	1.0	{0:24720.3,1:7313.9}	18504.62308

- Now that we have the output and the right formats, we need to find the minimum distance as discussed in the algorithm earlier. Scalding provides convenient `grouping` functions. First, we group the input rows on the `user` column and sort the `distance` column in the ascending order. This is `Reduce` operation.

```
import org.apache.mahout.common.distance.EuclideanDistanceMeasure
import org.apache.mahout.math.{NamedVector, DenseVector, VectorWritable}
import org.apache.mahout.clustering.iterator.ClusterWritable
import cascading.pipe.joiner._
import com.twitter.scalding._
class InputJob(args: Args) extends Job(args) {
val userVectors = WritableSequenceFile[Text, VectorWritable]
                (mahout_vectors, ('user , 'vector))
    .crossWithTiny(clusterCenter)
        .map(('center, 'vector) -> 'distance) {
        x:(DenseVector, VectorWritable) =>
                (new EuclideanDistanceMeasure()).distance(x._1,
                                                    x._2.get)
        }
        .map(('user, 'vector) -> ('user, 'vector)){
            x : (Text, VectorWritable) =>
                    (x._1.toString.toDouble, x._2.toString)
        }
        .groupBy(('user)){
        _.sortBy('distance)
        }
        .write(Tsv("Sorted.tsv"))
```

- Second, we take the sorted rows and take the first row of grouping, as it equates to the combination of input vector and centroid with minimum distance. Its API's like these that put Scalding at the top of the Highly productive hadoop wrapper frameworks.

Scalding at the top of the Highly

```
import org.apache.mahout.common.distance.EuclideanDistanceMeasure
import org.apache.mahout.math.{NamedVector, DenseVector, VectorWritable}
import org.apache.mahout.clustering.iterator.ClusterWritable
```

```
import cascading.pipe.joiner._
import com.twitter.scalding._
class InputJob(args: Args) extends Job(args) {
val userVectors  = WritableSequenceFile[Text, VectorWritable]
        (mahout_vectors, ('user , 'vector))
    .crossWithTiny(clusterCenter)
    .map(('center, 'vector) -> 'distance) {
        x:(DenseVector, VectorWritable) =>
          (new EuclideanDistanceMeasure()).distance(x._1,
                                                    x._2.get)
    }
    .map(('user, 'vector) -> ('user, 'vector)){
        x : (Text, VectorWritable) =>
          (x._1.toString.toDouble, x._2.toString)
    }
    .groupBy(('user)){
        _.sortBy('distance)
    }
    .groupBy(('user)){
        _.take(1)
    }
    .write(Tsv("Output.tsv"))
```

Step 10: Now that all the required input vectors, their corresponding Centroids, and associated classes are computed, you can see that output needs some cleaning up so that the output can be used for generating a graph. By cleaning up, the vectorized column needs to split to (x,y) tuples for both the input vector and centroids. You either can run a simple script to clean the code, or use scalding itself.

Using Scalding we can do this is two simple steps, though this not required and not in anyway needed to perform the KMeans task.

- First we perform a join, OuterJoin, the input dataset to the table 5.11 output. We join against the user and id fields.

```
import org.apache.mahout.common.distance.EuclideanDistanceMeasure
import org.apache.mahout.math.{NamedVector, DenseVector, VectorWritable}
import org.apache.mahout.clustering.iterator.ClusterWritable
import cascading.pipe.joiner._
import com.twitter.scalding._
class InputJob(args: Args) extends Job(args) {
val userVectors  = WritableSequenceFile[Text, VectorWritable]
        (mahout_vectors, ('user , 'vector))
    .crossWithTiny(clusterCenter)
    .map(('center, 'vector) -> 'distance) {
        x:(DenseVector, VectorWritable) =>
          (new EuclideanDistanceMeasure()).distance(x._1,
                                                    x._2.get)
    }
    .map(('user, 'vector) -> ('user, 'vector)){
        x : (Text, VectorWritable) =>
          (x._1.toString.toDouble, x._2.toString)
    }
    .groupBy(('user)){
        _.sortBy('distance)
    }
    .groupBy(('user)){
```

```
            _.take(1)
        }
        .joinWithSmaller('user -> 'id, userFeatures,
            joiner = new OuterJoin)
        .write(Tsv("cleaninput.tsv"))
```

- Second we join, Outerjoin, the collection from the previous step with the kmeans selected centroids. Since we performed an OutJoin, we have all the fields from the both the pipes/ collections. We can choose to project only the ones we need to graph, that is the input(x,y), ClusterId, Centroid(x,y).

```
import org.apache.mahout.common.distance.EuclideanDistanceMeasure
import org.apache.mahout.math.{NamedVector, DenseVector, VectorWritable}
import org.apache.mahout.clustering.iterator.ClusterWritable
import cascading.pipe.joiner._
import com.twitter.scalding._
class InputJob(args: Args) extends Job(args) {
val userVectors = WritableSequenceFile[Text, VectorWritable]
            (mahout_vectors, ('user , 'vector))
    .crossWithTiny(clusterCenter)
    .map(('center, 'vector) -> 'distance) {
        x:(DenseVector, VectorWritable) =>
        (new EuclideanDistanceMeasure()).distance(x._1,
                                                   x._2.get)
    }
    .map(('user, 'vector) -> ('user, 'vector)){
        x : (Text, VectorWritable) =>
            (x._1.toString.toDouble, x._2.toString)
    }
    .groupBy(('user)){
        _.sortBy('distance)
    }
    .groupBy(('user)){
        _.take(1)
    }
    .joinWithSmaller('user -> 'id, userFeatures,
            joiner = new OuterJoin)
    .joinWithSmaller('cid -> 'cxid, selectedclusters,
            joiner = new OuterJoin)
    .project('x, 'y, 'cid, 'cxx, 'cxy)
    .write(Tsv("final.tsv"))
}
```

The output of the above code looks like this:

Table 5.12

Size	Revenue	Class	Size(centroid)	Revenue(centroid)
6882	3758	0.0	7810.2727	4470.0
9916	5478	0.0	7810.2727	4470.0
9368	3521	0.0	7810.2727	4470.0
6386	4809	0.0	7810.2727	4470.0
25670	6921	1.0	24720.3	7313.9
28993	7852	2.0	24720.3	7313.9
20857	8833	3.0	24720.3	7313.9

The graph is drawn using simple R code and shown in Figure 5.4

```
#!/usr/bin/env Rscript
library(ggplot2)
x <- read.csv("final.tsv", header=FALSE, sep="\t")
names(x) <- c("size", "revenue", "class", "centroidsx",
                                           "centroidsy")

x$class <- factor(x$class)
x$classcentroids <- factor(x$centroidsx)
ggplot() +
    geom_point(data= x, aes(x=size, y=revenue,
                            color=class), size=4) +
    geom_point(data= x, aes(x=centroidsx, y=centroidsy),
                            shape=15, size=4) +
    ggtitle("Centroids After the KMeans Algorithm") +
    ggsave(file="graph.png", width=10, height=7)
```

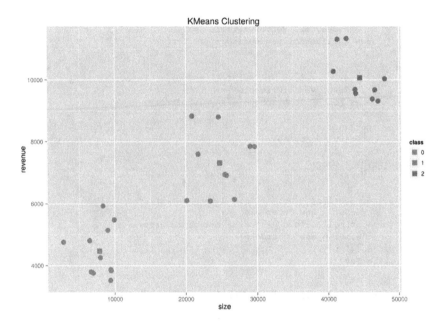

Fig. 5.4: K-Means Clustering output

Problems

5.1. Implement the above K-means algorithm given by the Apache Mahout library, using Spark and verify the results. Use the Spark Programming guide explained in Chapter 4. The complete source code is available at
`https://github.com/4nil/hpdc-scalding-spark`

References

1. StatStoft, Inc. *Data Mining Techniques*
   http://www.obgyn.cam.ac.uk/cam-only/statsbook/stdatmin.html
2. Jure Leskovec, Anand Rajaraman, Jeffrey D. Ullman,*Mining of Massive Datasets*, 2010
3. Mohammed J. Zaki, Wagner Meira JR. Ullman,*Data Mining and Analysis: Fundamental Concepts and Algorithms* , 2014
4. John. McCullok, *Step By Step K-Means*
   http://mnemstudio.org/clustering-k-means-example-1.htm
5. *NonLinear Dimensionality Reduction*
   http://en.wikipedia.org/wiki/Nonlinear_dimensionality_reduction
6. *Principal Componet Analysis*
   http://en.wikipedia.org/wiki/Principal_component_analysis
7. Rice, Stephen V., Nagy, George, Nartker, Thomas A. *Optical Character Recognition*, Springer, 1999
8. Han, J. and Pei, J. 2000. *Mining frequent patterns by pattern growth: Methodology and implications.* SIGKDD Explorations Newsletter 2, 2, 1420.
9. Han, Jiawei, and Micheline Kamber. *Data Mining, Southeast Asia Edition: Concepts and Techniques.* Morgan kaufmann, 2006.
10. Fayyad, Usama M., et al. "Advances in knowledge discovery and data mining." (1996).
11. Berkhin, Pavel. "A survey of clustering data mining techniques." Grouping multidimensional data. Springer Berlin Heidelberg, 2006. 25-71.
12. Weiss, Sholom M. Predictive data mining: a practical guide. Morgan Kaufmann, 1998.
13. Witten, Ian H., and Eibe Frank. Data Mining: Practical machine learning tools and techniques. Morgan Kaufmann, 2005.
14. Park, Byung-Hoon, and Hillol Kargupta. "Distributed data mining: Algorithms, systems, and applications." (2002).
15. Elavarasi, S. Anitha, J. Akilandeswari, and B. Sathiyabhama. "A survey on partition clustering algorithms." International Journal of Enterprise Computing and Business Systems 1.1 (2011).
16. Johnson, Stephen C. "Hierarchical clustering schemes." Psychometrika 32.3 (1967): 241-254.
17. Rayner Alfred, 2008, *A Data Summarisation Approach toKnowledge Discovery*, Thesis, Univeristy of York.

Chapter 6

Case Study II: Data Classification using Scalding and Spark

It is important to characterize learning problems depending on the type of data they use. Knowledge about the data is very important as similar learning techniques can be applied to similar data types. For example, Natural Language Processing and Bioinformatics use very similar tools for strings for natural language text and DNA sequences. The most basic type of data entities are **Vectors** . For example, an insurance corporation may want a vector of patient details like blood pressure, heart rate, height, weight, cholesterol, smoking status, and gender to infer the patients life expectancy. A farmer might be interested in determining the ripeness of the fruit based on a vector of size, weight, and spectral data. An electrical engineer may want to find dependency between voltage and current. A search engine might want to find a vector of counts which describe the frequency of words.

Handling scales and units of different attributes that vary widely in vectors can be challenging. For instance, weight can be measured in kilograms, pounds, grams, tons, stone ,etc. all of which are just multiplicative changes. Similarly, when representing temperature depending on the unit used Celsius, Kelvin, or Fahrenheit a class of transformations are available. One possible way to accommodate these scenarios is to normalize the data.

In some cases vectors can contain variable number of features. For instance, a physician can choose the tests, from a collection, to perform depending on the health status of the patient. To accommodate these **Lists** are used.

Consider a dataset that provides the toxic nature of mushrooms. It is desirable to infer the toxicity of new mushrooms based on the previous data and its chemical compounds. The toxicity of the mushroom may be identified by one or two compounds in the mushroom. **Sets** are used to infer the properties of a collection of features whose composition can vary significantly.

Matrices are used extensively in representing pairwise relationships. In collaborative filtering applications, matrix rows are user instances while columns are corresponding products. A combination of row and column could provide the rating of

© Springer International Publishing Switzerland 2015
K.G. Srinivasa and A.K. Muppalla, *Guide to High Performance Distributed Computing*, Computer Communications and Networks,
DOI 10.1007/978-3-319-13497-0_6

the user about the product and the matrix allows of null ratings. **Image** data can also be represented as two-dimensional arrays of number, in matrices. This representation is very crude as they also contain lines, shapes, and multiresolution structure. Downsampling an image leads to an object with similar image characteristics. There are many tools and libraries in computer vision and psycho-optics that manipulate images. **Video** is nothing but images with temporal dimensions. Videos can be represented as 3D arrays. Good algorithms accommodate the temporal characteristics of the images.

Relations between collections of objects can be expressed using **Trees and Graphs**. For example, the structure of the website *www.dmoz.org* can be represented as a Tree, where the topics are the nodes, they become increasingly refined as we traverse from the root node to the leaves: (Arts $\longrightarrow$ Animation $\longrightarrow$ Anime $\longrightarrow$ General Fan Pages $\longrightarrow$ Official Sites). In the case of gene sequencing the relationships take the form of a directed acyclic graph. Both these examples represent that the observations are vertices of a tree or graph. There are instances where the graphs are themselves observations, like in the case of call graph of a computer program, protein–protein interaction.

In the case of Bioinformatics and Natural Language Processing, **Strings** are very common. When modeling a topic structure in a document, when locating names of people and organizations in a text, Strings are the common inputs for many estimations. They are also popular output forms, while rendering a document summary, translating a document, etc.

The most common data types are **Compound Structure** which are a combination of simpler data types like Strings, Images, Tables which in turn contain numbers, lists, etc. Good statistical model takes into account these structures to build flexible models.

6.1 Classification

Development of Templates has significantly improved deployment rates of machine learning policies on new problems. The range of learning problem is quite large and growing constantly.

Binary Classification is one of the most frequently encountered problems and studied extensively over the years which has led to several algorithmic and theoretical breakthroughs. In its basic form: given x from a domain X estimate the value of an associated binary random variable $y \in \{\pm 1\}$ will assume. A simple example, given a collection of pictures of apples and oranges, predict whether the object is an apple or an orange. The application ranges from finding predicting load defaulters,

Spam, or Non-Spam email detection. There are several variants in binary classification estimation:

- A sequence of (x_i, y_i) pairs for which y_i is instantaneously estimated . This is commonly referred to as online learning.

- Observing a collection $\mathbf{X} := \{x_1, ...x_m\}$ and $\mathbf{Y} := \{y_1, ...y_m\}$ of pairs (x_i, y_i) which are then used to estimate y for a set of $X' = \{x_1', ..., x_m'\}$. This is commonly referred to as batch learning. Knowing X' already at the time of constructing the model is commonly referred to as transduction.

- Choosing $\mathbf{X}$ for the purpose of model building is known as active learning.

- Full information about $\mathbf{X}$ may not be available, e.g., some of the coordinates of the x_i might be missing, leads to the estimation of missing variables

- Covariate shift correction is done when two sets $\mathbf{X}$ and X' come from different data sources.

- Co-training is done when observations are collected from two problems which are somehow related.

- Depending on the type of error, estimation mistakes are penalized, e.g., when trying to distinguish diamonds from rocks, a very asymmetric loss applies.

Multi-Class Classification is an extension of Binary Classification. The main difference is that y may have a range of values: $y \in 1, 2, 3, ..., n$. For example, classifying documents according to the language was written in (English, French, German, Spanish, ...). See figure 6.1. The cost of error is heavily related to the type of error made. For example, in a scenario of classifying cancer patients, there is a huge difference in misclassifying an early stage cancer as healthy (patient is likely to die) or as advanced stage of cancer.

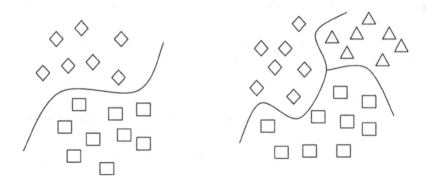

Fig. 6.1: Left: binary classification. Right: 3-class classification. Note that the latter case has much more degree for ambiguity. For instance, being able to distinguish diamonds from squares may not suffice to identify either of them correctly, the triangles need to be differentiated too

Structure Estimation: It takes multi-class classifying beyond simple means, as the labels for classification contain some structure which can be used in estimation. In collaborative filtering of ranking web pages, y might not be a path but a permutation when matching objects. Each of these problems has its own set of properties depending on which is found admissible.

6.2 Probability Theory

Machine Learning has wide range of applications. In order to understand these areas of applications there is a need to define a problem concisely. This section deals with a general overview of probability theory.

6.2.1 Random Variables

When a dice is cast, what are the chances it shows 1? If this dice is fair then there are six outcomes $\mathbf{X} = \{1,2,3,4,5,6\}$ which are equally likely to occur, hence the 1 would occur roughly once every 6 cases. Probability theory helps in modeling the uncertainty in the outcome of such experiments. Formally, the probability of state 1 in the previous example is $\frac{1}{6}$. In many cases the outcomes are numerical and can be handled easily, while others may not be numerical, e.g., observing heads or tails when a coin is tossed. In these nonnumeric cases the outcomes are mapped to numerical values. This is done using a *random variable* . For example, according to the previous example of a coin toss, a random variable X can be used that can take on a value of +1 if heads and -1 if tails. The notational convention of representing random variables is to use uppercase letters, e.g., X, Y, etc, and lowercase letters, e.g., x, y, etc, to denote their values.

6.2.2 Distributions

A random variable can be characterized by associating probabilities with the values it can take. If the random variable is discrete, that is it can accommodate finite number of values, then this probability assignment is called a *probability mass function* or PMF for short. A PMF by definition must be nonnegative and sum to one. For example, when a fair coin is tossed, heads and tails are equally likely and the random variable takes on the values +1 and -1 with probability 0.5. This can be formally written as:

$$Pr(X = +1) = 0.5 \text{ and } Pr(X = -1) = 0.5$$

A slightly informal notation is $p(x) := Pr(X = x)$.

Assignment of probabilities to a continuous random variable results in a *probability density function* , PDF for short. Like the PMF, the PDF must also be nonnegative and sum to one.

Uniform Distribution is given by the equation 6.2.2:

$$p(x) = \begin{cases} \frac{1}{b-a} & \text{if } x \in [a,b] \\ 0 & \text{otherwise} \end{cases}$$

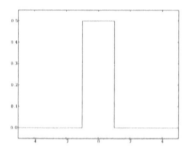 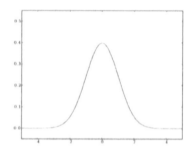

Fig. 6.2: Two common densities. Left: uniform distribution over the interval [1, 1]. Right: Normal distribution with zero mean and unit variance.

Gaussian distribution (also called normal distribution) is given by the equation 6.2.2.

$$p(x) = \frac{1}{\sqrt{2\pi\sigma^2}} exp\left(-\frac{(x-\mu)^2}{2\sigma^2}\right)$$

An integral over a PDF is termed as cumulative distribution function (CDF).

$$F_X(x) = \int_{-\infty}^{x} f_X(t)dt.$$

6.2.3 Mean and Variance

Determining the expected value of a random variable is commonly sought after. For instance, determining the typical values of voltages while measuring device voltage, determining expected height of a child before administering a growth hormone. Expectations and related entities are defined as follows:

We define *mean* of a random variable **X** as

$$\mathbb{E}[X] := \int xdp(x)$$

more generally, if $f : \mathbb{R} \longrightarrow \mathbb{R}$ is a function, then f(X) is a also a random variable. The mean of this function is given by:

$$\mathbb{E}[f(X)] := \int f(x)dp(x)$$

Whenever X is discrete random variable the integral can replaced by a summation:

$$\mathbb{E}[X] = \sum_{x} xp(x)$$

The dice has equal probabilities for all 6 possible outcomes. The mean translates to — (1+2+3+4+5+6)/6 = 3.5. The mean is used to estimate losses and gains. For example, in a stock market the expected value of a share over a year time is estimated. In addition to this, estimating the investment risk is also equally important. That is, what is the likelihood the value of the investment deviates from the expected outcome. To quantify the risk inherent of a random variable *variance* is used:

We define variance of random variable X as:

$$Var[X] := \mathbb{E}\left[(X - E[X])^2\right]$$

as before, if f: $\mathbb{R} \longrightarrow \mathbb{R}$, then the variance of f(X) is given by:

$$Var[f(X)] := \mathbb{E}\left[(f(X) - E[f(X)])^2\right]$$

The variance measures the degree by which function f (X) deviates from its expectation. An upper bound on the variance guarantees the probability lies within the expected value. This is one of the reasons for why the variance is measure of risk. Random variable in terms of *standard deviation* is defined as a square root of variance:

$$\sigma = \sqrt{\mathbb{E}[X^{\nu}] - (\mathbb{E}[X])^{\nu}}$$

6.3 Naive Bayes

It is an assumption with naive Bayes classifier that the value of a feature is independent of the presence or absence of other features. For example, a fruit can be considered an apple if it is red, round, about $3''$ in diameter. This classifier assumes that each of these features contribute independently to the probability of this fruit being an apple, regardless of other features.

In a supervised learning setup a naive Bayes classifier can be trained to be very effective. Maximum likelihood method is used to estimate the parameters of naive Bayes.

Despite their oversimplified assumptions, naive Bayes classifiers have proven worthy in solving many real-world problems. An advantage of naive bayes is that it requires significantly less data for classification. Since we use independent variables, only variances for each class is determined and not the entire covariance matrix.

6.3.1 Probabilty Model

The probability model for a classifier is given by a conditional model over a dependent variable C on several feature variables F_1 through F_n. When the number of feature variables is large then using probability tables to base this model is infeasible. We reformulate the model to make it more tractable:

$$p(C|F_1, ..., F_n)$$

Mathematically, Bayes theorem provides relationship between the probabilities of A and B, P(A) and P(B), and conditional probabilities of A given B and B given A, P(A|B) and P(B|A). The common form is:

$$P(A|B) = \frac{P(B|A)P(A)}{P(B)}$$

According to Bayesian interpretation, probability measures degree of belief. Bayes theorem links the belief metric with a hypothesis before and after account for evidence. For example, if the hypothesis is that a coin toss is twice as likely to land on heads as compared to tails. The belief degree is initially set at 50%, then the coin is tossed multiple times to evidence which may rise to 70% if the evidence matches the hypothesis.

For Hypothesis A and evidence B,

- P(A), the prior, is the initial degree of belief in A.

- P(A|B), the posterior, is the degree of belief given B has occurred

- the quotient P(B|A)/P(B) represents the support, B provides for A.

The probability model for dependent variable C and features F_1 through F_n is given by:

$$p(C|F_1, ...F_n) = \frac{p(C)p(F_1, ..., F_n|C)}{p(F_1, ..., F_n)}$$

In plain English, using Bayesian Probability terminology, the above equation can be written as

$$posterior = \frac{classprior \times liklihood}{evidence}$$

The emphasis is on numerator of the equation, because the denominator is independent of C and feature variables F_i are given, and it is effectively constant. The numerator is equivalent to the joint probability model:

$$p(C, F_1, ..., F_n)$$

which can be rewritten as follows, using the chain rule for repeated applications of
the definition of conditional probability:

$$
\begin{aligned}
p(C, F_1, \ldots, F_n) &= p(C)p(F_1, \ldots, F_n|C) \\
&= p(C)(F_1|C)p(F_2, \ldots F_n|C, F_1) \\
&= p(C)(F_1|C)p(F_2|C, F_1)p(F_3, \ldots F_n|C, F_1, F_2)
\end{aligned}
$$

The "naive" part of the theorem is in play: Each feature contributes independently
to the probability of the whole, that F_i is conditionally independent of every other
feature F_j for $j \neq i$ given the category C. This results in:

$$
\begin{aligned}
p(C|F_1, \ldots, F_n) &\propto p(C, F_1, \ldots, F_n) \\
&\propto p(C)p(F_1|C)p(F_2|C)p(F_3|C) \ldots \\
&\propto p(C)\prod_{i=1}^{n} p(F_i|C)
\end{aligned}
$$

This means that under the above independence assumptions, the conditional distribution over the class variable C is:

$$
p(C|F_1, \ldots, F_n) = \frac{1}{Z}p(C)\prod_{i=1}^{n} p(F_i|C)
$$

where the evidence $Z = p(F_1, \ldots, F_n)$ is a scaling factor dependent only on $F_1, \ldots, F_n$,
that is, a constant if the values of the feature variables are known.

Constructing a classifier from the probability model:

So far the naive Bayes probability model has been derived. The classifier combines
the model with a decision metric. A common approach to picking the rule is to use
the most probable hypothesis. The Bayes classifier function classify is defined as
follows:

$$
\text{classify}(f_1, \ldots, f_n) = \underset{c}{\text{argmax}}\, p(C = c)\prod_{i=1}^{n} p(F_i = f_i|C = c).
$$

6.3.2 Parameter Estimation and Event Models

The relative frequencies of all model parameters like class priors, feature probability distribution can be determined from the training set. These are called maximum likelihood estimates. The *class prior* can either be calculated assuming classes with equal probabilities, that is `priors = 1 / (number of classes)` or by estimating probability of the class from the training set, that is `class prior = (number of samples in class) / (total number of samples)`. A distribution is assumed or a nonparametric model for the features from the training set is assumed to estimate the parameters.

The event model of the naive Bayes classifier is the assumptions on feature distributions. There are two commonly used event models:

Gaussian Naive Bayes: If the data is continuous, the assumption is that the data follows a *Gaussian* distribution. For example, consider a training set with a continuous attribute x. First, the data is segmented by class, followed by the computation of *mean* and *variance*. The mean of the values of x with class c is denoted by μ_c and the variance of the values in x associated with class c is denoted by σ_c^2. Then, the probability density of some value given a class, P(x=v |c), can be computed by plugging v into the equation for a Normal distribution parameterized by μ_c and σ_c^2. That is,

$$p(x = v|c) = \frac{1}{\sqrt{2\pi\sigma_c^2}} e^{-\frac{(v-\mu_c)^2}{2\sigma_c^2}}$$

Binning is used to handle continuous values to discretize the features, obtaining a new set of *Bernoulli* distributed features. This type of distribution is preferable when there is small of training data or the data is precisely distributed. Binning is preferred in places where there is large amount of data, as it will learn to fit the data. The discretization method is normally preferred as naive Bayes implementation deals with large amount of data.

Multinomial Naive Bayes: In a multinomial event model, the feature vectors represent the event frequencies generated by a multinomial $(p_1,\ldots,p_n)$. Here p_i is the probability of even i. This type of model is normally applied to document classification; term frequencies are feature values, generated by a multinomial that produces a collection of words. The likelihood of observing a feature vector (histogram) F is given by

$$p(F|C) = \frac{(\sum_i F_i)!}{\prod_i F_i!} \prod_i p_i^{F_i}$$

The multinomial naive Bayes classifier becomes a linear classifier when expressed in log space:

$$\log p(C|F) = \log \left(p(C) \prod_{i=1}^{n} p(F_i|C) \right)$$

$$= \log p(C) + \sum_{i=1}^{n} \log p(F_i|C)$$

$$= b + \mathbf{w}_C^T \mathbf{F}$$

where $b = \log p(C)$ and $w_{Ci} = \log p(F_i|C)$

Despite the ambitious assumptions of feature independence which are often inaccurate, the classifier has several properties that make it very useful in practice. The class conditional feature distributions can independently estimated as a one-dimensional distribution due to this assumption. The curse of dimensionality problems are alleviated, such as datasets that scale exponentially with features. In many applications an approximate classification is sufficient and naive Bayes provides a good estimate for the correct class even though it fails at being accurate. For example, the naive Bayes classifier makes pick the correct decision rule provided the correct class is more probable. This is true irrespective of estimated probability value. This way the classifier can be robust.

6.3.3 Example

Sex Classification : The goal is to classify the given data point (person) is a male or a female, using the features describing the data point. These features include height, weight, and foot size.

Training :

Sex	Height	Weight	Foot Size
male	6	180	12
male	5.92 (5'11")	190	11
male	5.58 (5'7")	170	12
male	5.92 (5'11")	165	10
female	5	100	6
female	5.5 (5'6")	150	8
female	5.42 (5'5")	130	7
female	5.75 (5'9")	150	9

A Gaussian distribution-based classifier will have:

Sex	Mean(Height)	Mean(weight)	Mean(footsize)
male	5.855	176.25	11.25
female	5.4175	132.5	7.5

Sex	Variance(Height)	Variance(weight)	Variance(footsize)
male	3.5033e-02	1.2292e+02	9.1667e-01
female	9.7225e-02	5.5833e+02	1.6667e+00

Let us say we have classes which are equally likely, so P(male)= P(female) = 0.5. This prior probability distribution might be based on our knowledge of frequencies in the larger population, or on frequency in the training set.

Testing :

Below is a sample to be classified as a male or female.

Sex	Height	weight	footsize
sample	6	130	8

We wish to determine which posterior is greater, male or female. For the classification as male the posterior is given by

$$posterior(male) = \frac{P(male)\,p(height|male)\,p(weight|male)\,p(footsize|male)}{evidence}$$

For the classification as female the posterior is given by

$$posterior(female) \\ = \frac{P(female)\,p(height|female)\,p(weight|female)\,p(footsize|female)}{evidence}$$

The evidence (also termed normalizing constant) may be calculated as

$$evidence = P(male)p(height|male)p(weight|male)p(footsize|male)$$
$$+ P(female)p(height|female)p(weight|female)p(footsize|female)$$

However, given the sample the evidence is a constant and thus scales both posteriors equally. It therefore does not affect classification and can be ignored. We now determine the probability distribution for the sex of the sample.

$P(male) = 0.5$

$$p(height|male) = \frac{1}{\sqrt{2\pi\sigma^2}} exp\left(-\frac{(6-\mu)^2}{2\sigma^2}\right) \approx 1.5789$$

where $\mu = 5.855$ and $\sigma^2 = 3.5033 \cdot 10^{-2}$

are the parameters of normal distribution which have been previously determined from the training set. Note that a value greater than 1 is OK here it is a probability density rather than a probability, because height is a continuous variable.

$$p(weight|male) = 5.9881 \cdot 10^{-6}$$
$$p(footsize|male) = 1.3112 \cdot 10^{-3}$$
$$posteriornumerator(male) = theirproduct = 6.1984 \cdot 10^{-9}$$
$$P(female) = 0.5$$
$$p(height|female) = 2.2346 \cdot 10^{-1}$$
$$p(weight|female) = 1.6789 \cdot 10^{-2}$$
$$p(footsize|female) = 2.8669 \cdot 10^{-1}$$
$$posteriornumberator(femle) = theirproduct = 5.3778 \cdot 10^{-4}$$

Since posterior numerator is greater in the female case, we predict the sample is *female*.

6.4 Implementation of Naive Bayes Classifier

Here the objective is to implement the naive Bayes classifier using Scalding and Spark framework on the iris dataset.

Dataset The Iris dataset contains 150 instances, corresponding to three equally frequent species of iris plant (Iris setosa, Iris versicolour, and Iris virginica). One class is linearly separable from the other two; the latter are NOT linearly separable from each other.

Attributes/Features

- sepal Length in cm

- sepal width in cm

- petal length in cm

- petal width in cm

Sample Data Below is sample 20 rows of the dataset

id	class	sepallength	sepalwidth	petallength	petalwidth
0	0	5.0999999999999996	3.5	1.3999999999999999	0.20000000000000001
1	0	4.9000000000000004	3.0	1.3999999999999999	0.20000000000000001
2	0	4.7000000000000002	3.2000000000000002	1.3	0.20000000000000001
3	0	4.5999999999999996	3.1000000000000001	1.5	0.20000000000000001
4	0	5.0	3.6000000000000001	1.3999999999999999	0.20000000000000001
5	0	5.4000000000000004	3.8999999999999999	1.7	0.40000000000000002
6	0	4.5999999999999996	3.3999999999999999	1.3999999999999999	0.29999999999999999
7	0	5.0	3.3999999999999999	1.5	0.20000000000000001
8	0	4.4000000000000004	2.8999999999999999	1.3999999999999999	0.20000000000000001
9	0	4.9000000000000004	3.1000000000000001	1.5 0.10000000000000001	
10	0	5.4000000000000004	3.7000000000000002	1.5 0.20000000000000001	
11	0	4.7999999999999998	3.3999999999999999	1.6000000000000001	0.20000000000000001
12	0	4.7999999999999998	3.0	1.3999999999999999	0.10000000000000001
13	0	4.2999999999999998	3.0	1.1000000000000001	0.10000000000000001
14	0	5.7999999999999998	4.0	1.2	0.20000000000000001
15	0	5.7000000000000002	4.4000000000000004	1.5	0.40000000000000002
16	0	5.4000000000000004	3.8999999999999999	1.3	0.40000000000000002
17	0	5.0999999999999996	3.5	1.3999999999999999	0.29999999999999999
18	0	5.7000000000000002	3.7999999999999998	1.7	0.29999999999999999
19	0	5.0999999999999996	3.7999999999999998	1.5	0.29999999999999999
20	0	5.4000000000000004	3.3999999999999999	1.7	0.20000000000000001

Task Predicted attribute: class of iris plant from 3 classes:

- Iris Setosa
- Iris Versicolour
- Iris virginical

6.4.1 Scalding Implementation

Step 1: Initialize Scalding Job, here NBJob extends `Job` class that implements the necessary I/O formatting and all other hadoop configuration. Scala Programming language is used.

```
class NBJob(args: Args) extends Job(args) {

  // Code goes here

}
```

Step 2: Read Input using the fields based on API of scalding, we store the input as discussed earlier into the fields using `Tsv`

```
class NBJob(args: Args) extends Job(args) {
  val input = args("input")
  val output = args("output")

  val iris = Tsv(input, ('id, 'class,
  'sepalLength, 'sepalWidth, 'petalLength, 'petalWidth))
    .read
    .write(Tsv(output))
}
```

Step 3: Unpivot the fields, sepalLength , sepalWidth , petalLength , petalWidth to feature , score.

The scalding API demonstrates the productivity gain by providing simple abstractions to read/write data. Here `write(FORMAT)` is used to write out the dataset in `Tsv` (Tab Separated values) format.

```
class NBJob(args: Args) extends Job(args) {
  val input = args("input")
  val output = args("output")

  val iris = Tsv(input, ('id, 'class,
  'sepalLength, 'sepalWidth, 'petalLength, 'petalWidth))
    .read

  val irisMelted = iris
```

```
    .unpivot ((' sepalLength, ' sepalWidth,
    ' petalLength, ' petalWidth) -> (' feature, ' score))
    .write(Tsv(output))
}
```

Sample output:

id	class	feature	score
0	0	sepalLength	5.0999999999999996
0	0	sepalWidth	3.5
0	0	petalLength	1.3999999999999999
0	0	petalWidth	0.20000000000000001
1	0	sepalLength	4.9000000000000004
1	0	sepalWidth	3.0
1	0	petalLength	1.3999999999999999
1	0	petalWidth	0.20000000000000001
2	0	sepalLength	4.7000000000000002
2	0	sepalWidth	3.2000000000000002
2	0	petalLength	1.3
2	0	petalWidth	0.20000000000000001
3	0	sepalLength	4.5999999999999996
3	0	sepalWidth	3.1000000000000001
3	0	petalLength	1.5

Step 4: Choose the training dataset. From the dataset we choose the training set by filtering the rows whose id's are not divisible by 3, and discard the id column, i.e., every 3rd element is part of the training set.

```
class NBJob(args: Args) extends Job(args) {
  val input = args("input")
  val output = args("output")

  val iris = Tsv(input, ('id, 'class,
  'sepalLength, 'sepalWidth, 'petalLength, 'petalWidth))
    .read

  val irisMelted = iris
    .unpivot (('sepalLength, 'sepalWidth,
    'petalLength, 'petalWidth) -> ('feature, 'score))

}
    val irisTrain = irisMelted.filter('id){
    id: Int => (id % 3) != 0}.discard('id)
      .write(Tsv(output))
```

Sample Output: The Train dataset contains class, feature, score columns left.

class	feature	score
0	sepalLength	4.9000000000000004
0	sepalWidth	3.0
0	petalLength	1.3999999999999999
0	petalWidth	0.20000000000000001
0	sepalLength	4.7000000000000002
0	sepalWidth	3.2000000000000002
0	petalLength	1.3
0	petalWidth	0.20000000000000001
0	sepalLength	5.0
0	sepalWidth	3.6000000000000001
0	petalLength	1.3999999999999999
0	petalWidth	0.20000000000000001
0	sepalLength	5.4000000000000004
0	sepalWidth	3.8999999999999999
0	petalLength	1.7

Step 5: Now that the train dataset has been filtered and the remaining dataset can be used to test the model developed. Since we chose the row id's which were not divisible by 3 for train we chose the row id's which are divisible by 3 for test. Also, since the original dataset has the classes assigned to the features we need to remove the class column from the test dataset, to enable the model to classify it.

```
class NBJob(args: Args) extends Job(args) {

val input = args("input")
val output = args("output")

val iris = Tsv(input, ('id, 'class, 'sepalLength,
          'sepalWidth, 'petalLength, 'petalWidth))
.read

val irisMelted = iris
.unpivot(('sepalLength, 'sepalWidth, 'petalLength,
          'petalWidth) -> ('feature, 'score))

val irisTrain = irisMelted.filter('id)
            {id: Int => (id % 3) != 0}.discard('id)

val irisTest = irisMelted
  .filter('id){id: Int => (id % 3) ==0}
  .discard('class)
  .write(Tsv(output))

}
```

Step 6: Count the strength of each class from Train set.

```
class NBJob(args: Args) extends Job(args) {

val input = args("input")
```

```
val output = args("output")

val iris = Tsv(input, ('id, 'class, 'sepalLength,
          'sepalWidth, 'petalLength, 'petalWidth))
.read

val irisMelted = iris
.unpivot(('sepalLength, 'sepalWidth, 'petalLength,
          'petalWidth) -> ('feature, 'score))

val irisTrain = irisMelted.filter('id)
              {id: Int => (id % 3) != 0}.discard('id)

val irisTest = irisMelted
  .filter('id){id: Int => (id % 3) ==0}
  .discard('class)

val counts = irisTrain.groupBy('class) {
              _.size('classCount).reducers(10) }
    .write(Tsv(output))
  }
```

Output The class and count of the number of features each class has.

class	classcount
0	132
1	132
2	136

Step 7: Total class count

```
class NBJob(args: Args) extends Job(args) {

val input = args("input")
val output = args("output")

val iris = Tsv(input, ('id, 'class, 'sepalLength,
          'sepalWidth, 'petalLength, 'petalWidth))
.read

val irisMelted = iris
.unpivot(('sepalLength, 'sepalWidth, 'petalLength,
          'petalWidth) -> ('feature, 'score))

val irisTrain = irisMelted.filter('id)
              {id: Int => (id % 3) != 0}.discard('id)

val irisTest = irisMelted
  .filter('id){id: Int => (id % 3) ==0}
  .discard('class)
```

```
val counts = irisTrain.groupBy('class) {
              _.size('classCount).reducers(10) }

val totSum = counts.groupAll(_.sum[Double](
             'classCount -> 'totalCount))
         .write(Tsv(output))
}
```

Output The class and count of the number of features each class has.

totalcount
400

Step 8: Determine the ClassPrior / p(C), which is $\log(\frac{classcount}{totalcount})$. This is achieved by

- create a cross product of class, classcount, and totalcount.
- for each row calculate the classprior by $\log(\frac{classcount}{totalcount})$, using a mapTo function. The fields (class, classcount, totalcount) $\Longrightarrow$ (class, classPrior, classcount)
- discard the classcount

```
class NBJob(args: Args) extends Job(args) {

val input = args("input")
val output = args("output")

val iris = Tsv(input, ('id, 'class, 'sepalLength,
         'sepalWidth, 'petalLength, 'petalWidth))
.read

val irisMelted = iris
.unpivot(('sepalLength, 'sepalWidth, 'petalLength,
          'petalWidth) -> ('feature, 'score))

val irisTrain = irisMelted.filter('id)
              {id: Int => (id % 3) != 0}.discard('id)

val irisTest = irisMelted
 .filter('id){id: Int => (id % 3) ==0}
 .discard('class)

val counts = irisTrain.groupBy('class) {
              _.size('classCount).reducers(10) }

val totSum = counts.groupAll(_.sum[Double](
             'classCount -> 'totalCount))
```

```
                  .write(Tsv(output))

val prClass = counts
  .crossWithTiny(totSum)
  .mapTo(('class, 'classCount, 'totalCount) ->
            ('class, 'classPrior, 'classCount)) {
       x : (String, Double, Double) =>
                  (x._1, math.log(x._2 / x._3), x._2)
  }
  .discard('classCount)
  .write(Tsv(output))
  }
```

Output Class and ClassPrior

class	classprior
0	-1.1086626245216111
1	-1.1086626245216111
2	-1.0788096613719298

Step 9: According to Gaussian Naive Bayes implementation, we need to calculate the average(mean) and variance. Scalding provides ready-made solutions for this in the form of a function `sizeAveStdev`, which is a part of the grouping function.

```
group.sizeAveStdev(field, fields)
```

Sample Usage: Find the count of boys versus girls, their mean age, and standard deviation. The new pipe contains "sex", "count", "meanAge", and "stdevAge" fields. More in Chapter 4.

```
val demographics = people.groupBy('sex){
  _.sizeAveStdev('age->('count,'meanAge,'stdevAge))}
```

Apply the same concept to our problem. The Train set contains fields `class`, `feature`, `score`. We perform a groupby operation,

groupBy(feature, score) $\implies$ (featureclasssize, theta, sigma)

The groupby method also takes a reducers argument. We have used 10.

```
class NBJob(args: Args) extends Job(args) {

val input = args("input")
val output = args("output")

val iris = Tsv(input, ('id, 'class, 'sepalLength,
        'sepalWidth, 'petalLength, 'petalWidth))
  .read
```

```
val irisMelted = iris
.unpivot ((' sepalLength, 'sepalWidth, 'petalLength,
            'petalWidth) -> ('feature, 'score))

val irisTrain = irisMelted.filter('id)
                {id: Int => (id % 3) != 0}.discard('id)

val irisTest = irisMelted
  .filter('id){id: Int => (id % 3) ==0}
  .discard('class)

val counts = irisTrain.groupBy('class) {
              _.size('classCount).reducers(10) }

val totSum = counts.groupAll(_.sum[Double](
                'classCount -> 'totalCount))
                .write(Tsv(output))

val prClass =
  counts
      .crossWithTiny(totSum)
      .mapTo(('class, 'classCount, 'totalCount) ->
                  ('class, 'classPrior, 'classCount)) {
        x : (String, Double, Double) =>
                  (x._1, math.log(x._2 / x._3), x._2)
      }
      .discard('classCount)

val prFeatureClass = irisTrain
  .groupBy('feature, 'class) {
    _.sizeAveStdev('score ->
        ('featureClassSize, 'theta, 'sigma))
      .reducers(10)
  }
  .write(Tsv(output))
}
```

Step 10: Training the Model

Joins two pipes on a specified set of fields. Use this when pipe2 has fewer rows than pipe1. The Join declaration is as follows:

```
pipe1.joinWithSmaller(fields, pipe2)
```

Example: Here people is a large pipe with a *"birthCityId"* field. Join it against the smaller *"cities"* pipe, which contains an *"id"* field.

```
val peopleWithBirthplaces = people.joinWithSmaller
                  ('birthCityId -> 'id, cities)
```

Join on both *city.id* and *state.id*

```
val peopleWithBirthplaces = people.joinWithSmaller(
   ('birthCityId,'birthStateID)->('id,'StateID),cities)
```

Training can be further divided into:

- First Join operation between the `irisTrain` set and `prClass`, Class Prior, determined in the previous step, Table 6.1.

  Sample output:

Table 6.1: Join 1

Class	Feature	Score	classPrior
0	sepalLength	4.9000000000000004	-1.1086626245216111
0	sepalWidth	3.0	-1.1086626245216111
0	petalLength	1.3999999999999999	-1.1086626245216111
1	petalWidth	1.8	-1.1086626245216111
1	sepalLength	6.0999999999999996	-1.1086626245216111
1	sepalWidth	2.7999999999999998	-1.1086626245216111

- Second Join operation between the result of the pervious join `irisTrain` and `prFeatureClass` determined in the previous step, Table 6.2.

  Sample output:

Table 6.2: Join 2

Class	Feature	score	classPrior	FCSize	theta	sigma
0	petalLength	1.3999999	-1.10866111	33	1.45151512	0.1844114
0	petalLength	1.3	-1.10866111	33	1.45151512	0.1844114
0	petalLength	1.3999999	-1.10866111	33	1.45151512	0.1844114
1	petalLength	4.7000002	-1.10866111	33	4.29090091	0.45684435
1	petalLength	4.9000004	-1.10866111	33	4.29090091	0.4568443
1	petalLength	4.0	-1.10866111	33	4.29090091	0.4568443

```
class NBJob(args: Args) extends Job(args) {

val input = args("input")
val output = args("output")

val iris = Tsv(input, ('id, 'class, 'sepalLength,
          'sepalWidth, 'petalLength, 'petalWidth))
.read

val irisMelted = iris
.unpivot(('sepalLength, 'sepalWidth, 'petalLength,
```

```
            'petalWidth) -> ('feature, 'score))

val irisTrain = irisMelted.filter('id)
             {id: Int => (id % 3) != 0}.discard('id)

val irisTest = irisMelted
  .filter('id){id: Int => (id % 3) ==0}
  .discard('class)

val counts = irisTrain.groupBy('class) {
             _.size('classCount).reducers(10)  }

val totSum = counts.groupAll(_.sum[Double](
              'classCount -> 'totalCount))
          .write(Tsv(output))

val prClass =
  counts
      .crossWithTiny(totSum)
      .mapTo(('class, 'classCount, 'totalCount) ->
              ('class, 'classPrior, 'classCount)) {
        x : (String, Double, Double) =>
               (x._1, math.log(x._2 / x._3), x._2)
      }
      .discard('classCount)

val prFeatureClass =
  irisTrain
      .groupBy('feature, 'class) {
        _.sizeAveStdev('score ->
             ('featureClassSize, 'theta, 'sigma))
          .reducers(10)
      }

val model = irisTrain
  .joinWithSmaller('class ->
       'class, prClass, reducers=10)
  .joinWithSmaller(('class, 'feature) ->
        ('class, 'feature), prFeatureClass, reducers=10)
  .mapTo(('class, 'classPrior, 'feature,
          'featureClassSize, 'theta, 'sigma) ->
           ('class, 'feature, 'classPrior,
                     'theta, 'sigma)) {
     values : (String, Double, String,
                 Double, Double, Double) =>
     val (classId, classPrior, feature,
       featureClassSize, theta, sigma) = values
     (classId, feature, classPrior,
               theta, math.pow(sigma, 2))
   }

  .write(Tsv(output))
  }
```

Step 11: Join the Training set with the Model result on Test set. The function skewJoinWithSmaller is used to join data that has extreme skew. For example, consider a Twitter dataset, where two pipes of followers and genders, respectively, are joined. Since Twitter followers are not distributed evenly, some people have large number of followers, this data will be passed to a single reducer during computation which can lead to performance bottlenecks. The skew function helps in alleviating this problem.

```
def skewJoinWithSmaller(fs : (Fields, Fields),
        otherPipe : Pipe, sampleRate : Double = 0.001,
        reducers : Int = -1,
        replicator : SkewReplication =
                        SkewReplicationA())
```

This works as follows:

- First samples form the two pipes are taken with some probability and checked for frequency of join key.

- We use a replication strategy defined to replicate these join keys.

- Finally, we join the replicated pipes together.

sampleRate: This controls how often we sample from the left and right pipes when estimating key counts.

replicator: Algorithm for determining how much to replicate a join key in the left and right pipes.

Note: Since we do not set the replication counts, only inner joins are allowed. (Otherwise, replicated rows would stay replicated when there is no counterpart in the other pipe.)

```
class NBJob(args: Args) extends Job(args) {

val input = args("input")
val output = args("output")

val iris = Tsv(input, ('id, 'class, 'sepalLength,
        'sepalWidth, 'petalLength, 'petalWidth))
.read

val irisMelted = iris
.unpivot(('sepalLength, 'sepalWidth, 'petalLength,
        'petalWidth) -> ('feature, 'score))

val irisTrain = irisMelted.filter('id)
            {id: Int => (id % 3) != 0}.discard('id)

val irisTest = irisMelted
 .filter('id){id: Int => (id % 3) ==0}
 .discard('class)

val counts = irisTrain.groupBy('class) {
```

```scala
                    _.size('classCount).reducers(10) }

val totSum = counts.groupAll(_.sum[Double](
            'classCount -> 'totalCount))
        .write(Tsv(output))

val prClass =
 counts
    .crossWithTiny(totSum)
    .mapTo(('class, 'classCount, 'totalCount) ->
            ('class, 'classPrior, 'classCount)) {
      x : (String, Double, Double) =>
            (x._1, math.log(x._2 / x._3), x._2)
    }
    .discard('classCount)

val prFeatureClass =
 irisTrain
    .groupBy('feature, 'class) {
      _.sizeAveStdev('score ->
          ('featureClassSize, 'theta, 'sigma))
        .reducers(10)
    }

val model = irisTrain
 .joinWithSmaller('class ->
      'class, prClass, reducers=10)
 .joinWithSmaller(('class, 'feature) ->
        ('class, 'feature), prFeatureClass, reducers=10)
 .mapTo(('class, 'classPrior, 'feature,
          'featureClassSize, 'theta, 'sigma) ->
          ('class, 'feature, 'classPrior,
                      'theta, 'sigma)) {
     values : (String, Double, String,
                  Double, Double, Double) =>
     val (classId, classPrior, feature,
      featureClassSize, theta, sigma) = values
     (classId, feature, classPrior,
                  theta, math.pow(sigma, 2))
   }

val joined = irisTest
  .skewJoinWithSmaller('feature -> 'feature,
                          model, reducers=10)
  .write(Tsv(output))
}
```

Step 12: Apply the gaussian probablity and determine the class prediction

```scala
class NBJob(args: Args) extends Job(args) {

val input = args("input")
val output = args("output")

val iris = Tsv(input, ('id, 'class, 'sepalLength,
        'sepalWidth, 'petalLength, 'petalWidth))
.read

val irisMelted = iris
.unpivot(('sepalLength, 'sepalWidth, 'petalLength,
        'petalWidth) -> ('feature, 'score))

val irisTrain = irisMelted.filter('id)
            {id: Int => (id % 3) != 0}.discard('id)

val irisTest = irisMelted
 .filter('id){id: Int => (id % 3) ==0}
 .discard('class)

val counts = irisTrain.groupBy('class) {
            _.size('classCount).reducers(10) }

val totSum = counts.groupAll(_.sum[Double](
            'classCount -> 'totalCount))
            .write(Tsv(output))

val prClass =
 counts
    .crossWithTiny(totSum)
    .mapTo(('class, 'classCount, 'totalCount) ->
            ('class, 'classPrior, 'classCount)) {
      x : (String, Double, Double) =>
            (x._1, math.log(x._2 / x._3), x._2)
    }
    .discard('classCount)

val prFeatureClass =
 irisTrain
    .groupBy('feature, 'class) {
      _.sizeAveStdev('score ->
          ('featureClassSize, 'theta, 'sigma))
        .reducers(10)
    }

val model = irisTrain
 .joinWithSmaller('class ->
      'class, prClass, reducers=10)
 .joinWithSmaller(('class, 'feature) ->
        ('class, 'feature), prFeatureClass, reducers=10)
 .mapTo(('class, 'classPrior, 'feature,
        'featureClassSize, 'theta, 'sigma) ->
```

```scala
                ('class, 'feature, 'classPrior,
                          'theta, 'sigma)) {
       values : (String, Double, String,
                    Double, Double, Double) =>
       val (classId, classPrior, feature,
        featureClassSize, theta, sigma) = values
       (classId, feature, classPrior,
                    theta, math.pow(sigma, 2))
    }

val joined = irisTest
  .skewJoinWithSmaller('feature -> 'feature,
                          model, reducers=10)

def _gaussian_prob(theta : Double,
                    sigma : Double,
                    score : Double) : Double = {

val outside = -0.5 * math.log(math.Pi * sigma)
val expo = 0.5 * math.pow(score - theta, 2) / sigma
outside - expo
}

val result = joined
  .map(('theta, 'sigma, 'score) -> 'evidence) {
     values : (Double, Double, Double) =>
      _gaussian_prob(values._1, values._2, values._3)}
  .project('id, 'class, 'classPrior, 'evidence)
  .groupBy('id, 'class) {
      _.sum[Double]('evidence -> 'sumEvidence)
       .max('classPrior)
  }

  .mapTo(('id, 'class, 'classPrior, 'sumEvidence) ->
                       ('id, 'class, 'logLikelihood)) {
    values : (String, String, Double, Double) =>
      val (id, className, classPrior, sumEvidence) = values
      (id, className, classPrior + sumEvidence)
  }

  .groupBy('id) {
    _.sortBy('logLikelihood)
     .reverse
     .take(1)
     .reducers(10)
  }
  .rename(('id, 'class) -> ('id2, 'classPred))
  .write(Tsv(output))
}
```

Step 13: Map the class predictions

```scala
class NBJob(args: Args) extends Job(args) {

val input = args("input")
val output = args("output")

val iris = Tsv(input, ('id, 'class, 'sepalLength,
        'sepalWidth, 'petalLength, 'petalWidth))
.read

val irisMelted = iris
.unpivot(('sepalLength, 'sepalWidth, 'petalLength,
        'petalWidth) -> ('feature, 'score))

val irisTrain = irisMelted.filter('id)
            {id: Int => (id % 3) != 0}.discard('id)

val irisTest = irisMelted
 .filter('id){id: Int => (id % 3) ==0}
 .discard('class)

val counts = irisTrain.groupBy('class) {
            _.size('classCount).reducers(10) }

val totSum = counts.groupAll(_.sum[Double](
            'classCount -> 'totalCount))
        .write(Tsv(output))

val prClass =
 counts
    .crossWithTiny(totSum)
    .mapTo(('class, 'classCount, 'totalCount) ->
            ('class, 'classPrior, 'classCount)) {
        x : (String, Double, Double) =>
            (x._1, math.log(x._2 / x._3), x._2)
    }
    .discard('classCount)

val prFeatureClass =
 irisTrain
    .groupBy('feature, 'class) {
      _.sizeAveStdev('score ->
          ('featureClassSize, 'theta, 'sigma))
        .reducers(10)
    }

val model = irisTrain
 .joinWithSmaller('class ->
      'class, prClass, reducers=10)
 .joinWithSmaller(('class, 'feature) ->
      ('class, 'feature), prFeatureClass, reducers=10)
 .mapTo(('class, 'classPrior, 'feature,
        'featureClassSize, 'theta, 'sigma) ->
```

```scala
                   ('class, 'feature, 'classPrior,
                              'theta, 'sigma)) {
        values : (String, Double, String,
                      Double, Double, Double) =>
        val (classId, classPrior, feature,
         featureClassSize, theta, sigma) = values
        (classId, feature, classPrior,
                      theta, math.pow(sigma, 2))
    }

val joined = irisTest
  .skewJoinWithSmaller('feature -> 'feature,
                          model, reducers=10)

def _gaussian_prob(theta : Double,
                   sigma : Double,
                   score : Double) : Double = {


val outside = -0.5 * math.log(math.Pi * sigma)
val expo = 0.5 * math.pow(score - theta, 2) / sigma
outside - expo
}

val result = joined
  .map(('theta, 'sigma, 'score) -> 'evidence) {
    values : (Double, Double, Double) =>
      _gaussian_prob(values._1, values._2, values._3)}
  .project('id, 'class, 'classPrior, 'evidence)
  .groupBy('id, 'class) {
      _.sum[Double]('evidence -> 'sumEvidence)
      .max('classPrior)
  }

  .mapTo(('id, 'class, 'classPrior, 'sumEvidence) ->
                    ('id, 'class, 'logLikelihood)) {
    values : (String, String, Double, Double) =>
    val (id, className, classPrior, sumEvidence) = values
    (id, className, classPrior + sumEvidence)
  }

  .groupBy('id) {
      _.sortBy('logLikelihood)
      .reverse
      .take(1)
      .reducers(10)
  }
  .rename(('id, 'class) -> ('id2, 'classPred))

val results = iris
  .leftJoinWithTiny('id -> 'id2, result)
  .discard('id2)
  .map('classPred -> 'classPred) {x: String =>
                              Option(x).getOrElse("")}
```

```
  .project('id, 'class, 'classPred,
             'petalLength, 'petalWidth)
  .write(Tsv(output))
}
```

6.4.2 Results

The classification results of sepal length versus sepal width is shown in Figure 6.3, petal length versus petal width is shown in Figure 6.4, sepal length versus petal width is shown in Figure 6.5 and petal length versus sepal length is shown in Figure 6.6.

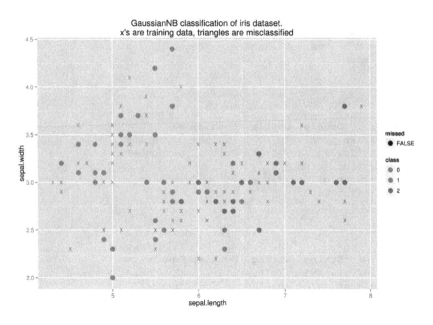

Fig. 6.3: Sepal Length versus Sepal Width

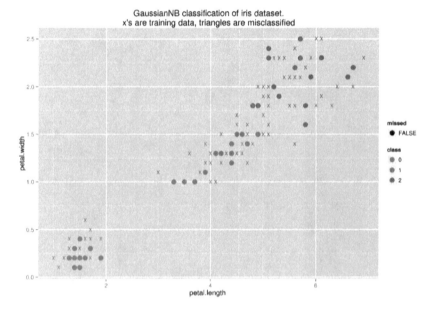

Fig. 6.4: Petal Length versus Petal Width

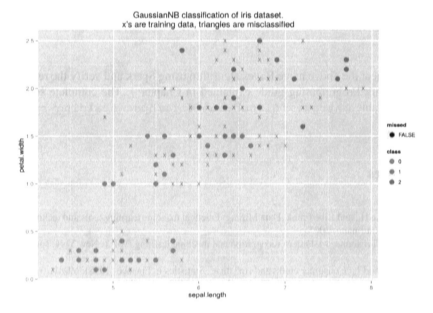

Fig. 6.5: Sepal Length versus Petal Width

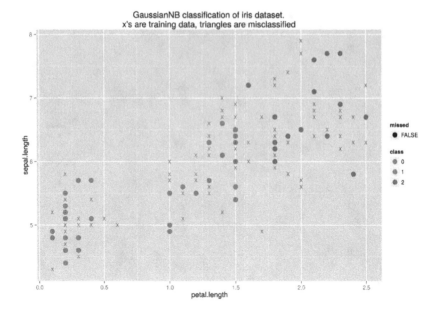

Fig. 6.6: Petal Length versus Sepal Length

Problems

6.1. Implement the above naive Bayes algorithm using Spark and verify the results. Use the Spark Programming guide explained in Chapter 4. The complete source code is available at https://github.com/4nil/hpdc-scalding-spark

References

1. Witten, Ian H., and Eibe Frank. Data Mining: Practical machine learning tools and techniques. Morgan Kaufmann, 2005.
2. Bishop, Christopher M. Pattern recognition and machine learning. Vol. 1. New York: springer, 2006.
3. Hand, David J. "Consumer credit and statistics." Statistics in Finance (1998): 69-81.
4. Alpaydin, Ethem. Introduction to machine learning. MIT press, 2004.
5. Bartlett, Marian Stewart, et al. "Recognizing facial expression: machine learning and application to spontaneous behavior." Computer Vision and Pattern Recognition, 2005. CVPR 2005. IEEE Computer Society Conference on. Vol. 2. IEEE, 2005.
6. Kononenko, Igor. "Machine learning for medical diagnosis: history, state of the art and perspective." Artificial Intelligence in medicine 23.1 (2001): 89-109.

7. Mitchell, Tom Michael. The discipline of machine learning. Carnegie Mellon University, School of Computer Science, Machine Learning Department, 2006.
8. Kotsiantis, Sotiris B., I. D. Zaharakis, and P. E. Pintelas. "Supervised machine learning: A review of classification techniques." (2007): 3-24.
9. Nguyen, Thuy TT, and Grenville Armitage. "A survey of techniques for internet traffic classification using machine learning." Communications Surveys & Tutorials, IEEE 10.4 (2008): 56-76.
10. Barto, Andrew G. Reinforcement learning: An introduction. MIT press, 1998.
11. Matthew Richardson, Amit Prakash, and Eric Brill. Beyond pagerank: machine learning for static ranking. In Les Carr, David De Roure, Arun Iyengar, Carole A. Goble, and Michael Dahlin, editors, WWW, pages 707-715. ACM, 2006.
12. Robert Bell and Yehuda Koren. Scalable collaborative filtering with jointly derived neighborhood interpolation weights. In Proceedings of the IEEE International Conference on Data Mining (ICDM), pages 43 52, 2007

Chapter 7

Case Study III: Regression Analysis using Scalding and Spark

Regression analysis is usually applied to prediction and forecasting with substantial overlap with the field of machine learning. The relationship between the dependent and independent variables is determined through regression and to explore different forms of these relationships. In certain circumstances where the assumptions are restricted, regression helps to infer a casual relationship. However, caution is advised as this can lead to illusions [1].

Investigating functional relationships among variables is simple using regression analysis. A real estate agent may assess the value of a home from some of the physical characteristics of a building and taxes(local, school, country) paid on the building. We can examine the relationship between cigarette consumption and age, sex, education, income, and price. This relationship can be expressed in the form of an equation or a model connecting the dependent and independent variable. With respect to the cigarette consumption example, the dependent variable *dependent variable is cigarette consumption - measured by the number of packs sold in a given state per capita income during a given year* and the independent variable is *various socio-economic and demographic variables like sex, age, etc.* In the real estate example, the dependent variable is the *price of the home* and the independent variable is the *physical characteristics of the home and the taxes paid on the building.* We denote the dependent variable as Y and the set of independent variables as $X_1, X_2, ..., X_p$, where p denotes the number of predictors. The relationship between these are represented by the model:

$$Y = f(X_1, X_2, ..., X_p) + \varepsilon$$

where ε is random error on the approximation. It accounts the errors in model fitting. The function $f(X_1, X_2, ..., X_p)$ describes the relationship between Y and $X_1, X_2, ..., X_p$ An example is the linear regression model:

$$Y = \beta_0 + \beta_1 X_1 + \beta_2 X_2 + \cdots + \beta_p X_p + \varepsilon$$

© Springer International Publishing Switzerland 2015
K.G. Srinivasa and A.K. Muppalla, *Guide to High Performance Distributed Computing*, Computer Communications and Networks, DOI 10.1007/978-3-319-13497-0_7

Where $\beta_0, \beta_1 X_1, ..., \beta_p X_p$ are called the regression parameters or coefficients that are to be determined from the data.

7.1 Steps in Regression Analysis

Regression analysis generally includes the following steps:

Step 1: Problems formulation is the first and perhaps the most important step to regression analysis because an ill-defined problem directly leads to wasted effort. It is important to determine the question to be answered to prevent choosing wrong features leading to wrong models. Consider an example where we determine if the employer is *discriminating* against female employees. To address this issue we collect data from the company such as *salary*, *qualifications*, and *sex* as features. We define *discrimination* in many ways, consider women are discriminated against if:

- Women are paid less compared to equally qualified Men?

- Women are more qualified than equally paid Men?

Feature Selection: We need to determine the *Dependent (Y)* and *Independent (X)* variables. For the first question, *Salary* is the dependent variable while *Qualification* and *Sex* are the independent variables. For the second question, *Qualification* is the dependent variable while *Salary* and *Sex* are the independent variables. It is important to note that the features are selected based on the question being answered.

Step 2: Data collection After the selection of variables for analysis, data from the environment is collected. Sometimes the data collected is under controlled settings where the parameters that are not of primary interest can be used as constants. More often the data collected is under nonexperimental conditions where there is little or no scope of control.

Observations as shown in the Table 7.1 consist of measurements for each of the potentially relevant data. The columns in this Table are features and each row represents an observation. The notation x_{ij} refers to the *i*th value of the *j*th variable. The first subscript is the observation number and the second is the variable number.

Each of the variables in the Table 7.1 can be classified as *quantitative* or *qualitative*. Some of the quantitative variable samples are : *price, number of bedrooms, age,* and *taxes*. Some of the qualitative variables are *neighborhood type, house style*, etc. In most cases, the variables are quantitative; however, the qualitative variables, if any, have to be used as *indicator* or *dummy* variables. In cases where the response variable is binary is called *logistic regression*. If some of the predictor variables are

Table 7.1: Data Notation used in Regression Analysis

Observation Number	Response	X_1	X_2	$\cdots$	X_2
1	y_1	x_{11}	x_{12}	$\cdots$	x_{1p}
2	y_2	x_{21}	x_{22}	$\cdots$	x_{2p}
3	y_3	x_{31}	x_{32}	$\cdots$	x_{3p}
4	y_4	x_{41}	x_{42}	$\cdots$	x_{4p}
5	y_5	x_{51}	x_{52}	$\cdots$	x_{5p}
.	.	.	.	.	.
.	.	.	.	.	.
.	.	.	.	.	.
n	y_n	x_{n1}	x_{n2}	$\cdots$	x_{np}

quantitative while others are qualitative, regression analysis in these cases is called the *analysis of covariance.*

Step 3: Model Specification Researchers can use their subjective and/or objective knowledge to ascertain the initial form of the model relating the dependent variable and the independent variable. The hypothesized model can then be either confirmed or refuted based on the analysis on the data. It is important to note that the model needs to be specified only in form and it can depend on unknown parameters. We need to select the form of the function $f(X_1, X_2, ..., X_p)$. This function can be classified into two types: linear and nonlinear. An example of a linear function is

$$Y = \beta_0 + \beta_1 X_1 + \varepsilon$$

while a nonlinear function is

$$y = \beta_0 + e^{\beta_1 X_1} + \varepsilon$$

Note: The term linear/ nonlinear here does not describe the relationship between Y and $X_1, X_2, ..., X_p$. It is related to the fact that the regression parameters enter the equation linearly / nonlinearly. Each of the following models are linear,

$$Y = \beta_0 + \beta_1 X + \beta_2 X^2 + \varepsilon$$
$$Y = \beta_0 + \beta_1 lnX + \varepsilon$$

because in each case the parameters enter linearly although the relationship between Y and X is nonlinear. This can be seen if the two models are re-expressed, respectively, as follows:

$$Y = \beta_0 + \beta_1 X_1 + \beta_2 X_2 + \varepsilon$$
$$Y = \beta_0 + \beta_1 X_1 + \varepsilon$$

where in the first equation we have $X_1 = X$ and $X_2 = X^2$ and in the second equation we have $X_1 = lnX$. The variables here are *re-expressed* or *transformed*.

Linearizable functions are nonlinear functions that can be transformed into linear functions. However, not all nonlinear functions are linearizable the following function can linearizable.

$$y = \beta_0 + e^{\beta_1 X_1} + \varepsilon$$

A regression equation containing only one predictor variable is called a simple regression equation. An equation containing more than one predictor variable is called a multiple regression equation. A simple regression example : the time to repair a machine is studied in relation to the number of components to be repaired. The independent variable here is the *number of components* and dependent variable is the *time to repair*. Multiple regression problem example: age-adjusted mortality rates prevailing in different geographic regions (response variables) by a large number of environmental and socioeconomic factors (predictor variables).

The independent variable need not always be a single variable, it can be a set of variables, $Y_1, Y_2, ..., Y_q$ say, which are thought to be related to the same set of predictor variables, $X_1, X_2, ..., X_p$. For example, data set presented in [3] consists of 148 healthy people. In this dataset 11 variables are selected, out of which 6 variables represent different types of measured sensory thresholds (e.g., vibration, hand and foot temperatures) and 5 baseline covariates (e.g., age, sex, height, and weight) that may effect on some or all of the six sensory thresholds. Here we have six response variables and five predictor variables. Further description of the data can be found in [3].

Univariate Regression deals with only one response variable; two or more response variables in regression are termed multivariate. Simple and multiple regression methods should be confused with univariate and multivariate regression analysis. The distinction between the two depends on the number of independent variables used, simple regression involved one independent variable while multiple regression modeling involved two or more [4].

Step 4: Model Fitting Now that we have collected the data and defined the model, we need to estimate the parameters of the model. This model estimation is also called *model fitting* . The commonly used method of estimation is *least squares* method. This method is desirable under certain assumptions. Other methods like *maximum likelihood*, *ridge*, and *principal components* can also be used if the least squares method assumptions do not hold.

The estimates of the regression parameters $\beta_0, \beta_1, ..., \beta_p$ are denoted by $\hat{\beta}_0, \hat{\beta}_1, ..., \hat{\beta}_p$ The estimated regression equation then becomes:

$$\hat{Y} = \hat{\beta}_0 + \hat{\beta}_1 X_1 + \hat{\beta}_2 X_2 + \cdots + \hat{\beta}_p X_p$$

The estimate of the parameter is denoted by a *hat $\hat{p}$* on top of a parameter.. The $\hat{Y}$ is called the fitted value. Using the input data, for each observation, we can compute the fitted value. For example, the *i*th fitted value $\hat{y}_i$ is:

$$\hat{y}_i = \hat{\beta}_0 + \hat{\beta}_1 x_i 1 + \hat{\beta}_2 x_i 2 + \cdots + \hat{\beta}_p x_i p, i = 1, 2, ..., n$$

where $x_{i1}, x_{i2}, ..., x_{ip}$ are the *i*th observation for p predictor variables. This equation can be used to predict the dependent variable for any values of the independent variables not in the observed data. In this the obtained Y is called the predicted value. The difference between *predicted* and *fitted* values is that the values used for the independent variable is one of the n observations in the data, while the predicted values are obtained for any value of the independent variable. The range of the independent variables is not recommended to be outside the range of our data during prediction. In scenarios where the independent variable values represent future values, then the predicted value is referred to as the *forecasted value* .

Step 5: Model Validation and criticism The validity of regression analysis and other such statistical methods depends on certain assumptions. The assumptions are usually about the data and the model. The accuracy of the model is dependent on the validity of these assumptions. To determine if the model assumptions hold, we need to address the following questions:

1. What are the assumptions?

2. How to validate these assumptions?

3. What can be done when more than one assumption does not hold?

Regression analysis is an iterative process in which the outputs are used to diagnose, validate, criticize, and modify the inputs. This process is repeated until a satisfactory output has been obtained. This output is an estimated model that satisfies the assumptions and fits the data reasonably well.

Objectives of Regression Analysis Determination of the regression equation is the most important product of the analysis. It is the summary of the relationship between the independent variable Y and a set dependent variables $X_1, X_2, ... X_n$. This equation can be used for many purposes: it may be used to evaluate the importance of individual predictors, to analyze the effects of the model by changing the values of the predictor variables, to forecast the response variable for a given set of independent variables.

Although the regression equation is a final product, there are a number of useful intermediate by-products. We also use these by-products to understand the relationships among variables in a certain environment. The task of regression analysis is to learn about the environment reflected by the data. We infer that what we uncover along the way to the formulation of the final equation is also important.

7.2 Implementation Details

Here the objective is to implement a linear regression model in Scalding and Spark frameworks to predict profits for a food truck based on population size of a city.

DataSet The dataset contains 97 data points: The first column is the population of a city and the second column is the profit of a food truck in that city. A negative value for profit indicates a loss.

- Column 0: Independent variable X, indicates the population of the city in the order of 10,000's

- Column 1: Dependent variable Y, indicates the size of the Profit of the food truck in the order of $10,000's

Using R to plot the Data:

```
#!/usr/bin/env Rscript
library(ggplot2)
x <- read.csv("in.txt", header=FALSE, sep=",")
names(x) <- c("x", "y")
ggplot() +
  geom_point(data=x, aes(x=x, y=y), size=4) +
  labs(x="Population in 10,000's",
      y ="Profit in  \$10,000's") +
  ggtitle("Food Truck Dataset") +
  ggsave(file="lrdataset.png", width=10, height=7)
```

The data distribution is shown in Figure 7.1.

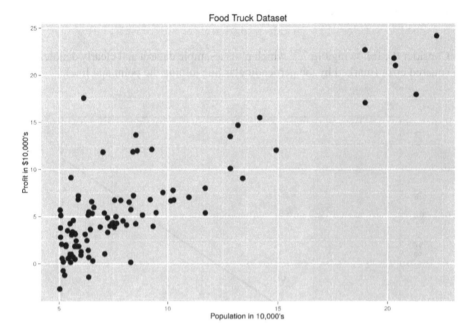

Fig. 7.1: Food Truck Dataset

Sample Dataset: A section of data is represented in Table 7.11

Table 7.2: Sample Dataset

Population (10,000's)	Profit ($10,000's)
6.1101	17.592
5.5277	9.1302
8.5186	13.662
7.0032	11.854
5.8598	6.8233
8.3829	11.886
7.4764	4.3483
8.5781	12
6.4862	6.5987
5.0546	3.8166

7.2.1 Linear Regression: Algebraic Method

Consider the following Fig 7.2, which plots a sample dataset and clearly denotes the squared error (marked by the perpendicular line joining the point and line).

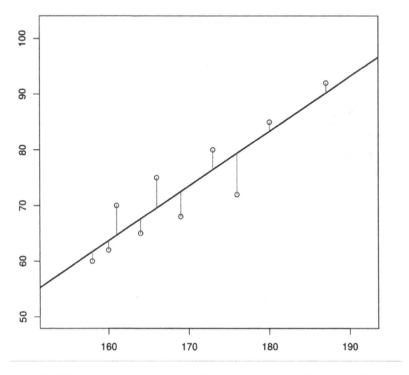

Fig. 7.2: Sample dataset showing the distance between the points and the line

Consider the points on the graph to be $(x_1,y_1), (x_2,y_2, (x_3,y_3,..., (x_n,y_n)$ and the equation of the line can expressed as $y = mx + b$. What we need to find is m and b so that it minimizes the Squared Error; as displayed on the graph, the squared errors are represented by the red lines.

The point where the red line segment meets the line $y = mx + b$ is nothing but $(x_i, mx_i + b)$; there the distance between points (x,y) and (x, mx+b) is $y - (mx+b)$.

This can be formulated as Squared Error(SE):

$$SE = (y_1 - (mx_1 + b))^2 + (y_1 - (mx_1 + b))^2 + ... + (y_1 - (mx_1 + b))^2$$

Expanding the above equation :-

$$SE = (y_1 - (mx_1 + b))^2 + (y_1 - (mx_1 + b))^2 + ... + (y_1 - (mx_1 + b))^2$$
$$= y_1^2 - 2y_1(mx_1 + b) + (mx_1 + b)^2 +$$
$$y_2^2 - 2y_2(mx_2 + b) + (mx_2 + b)^2 +$$

.
.

$$y_n^2 - 2y_n(mx_n + b) + (mx_n + b)^2$$

After a couple of algebraic manipulations and taking the mean, the equations look like this:

$$SE = n\overline{y^2} - 2mn\overline{xy} - 2bn\overline{y} + m^2 n\overline{x^2} + 2mbn\overline{x} + nb^2$$

Now to minimize the above equation and satisfy our original premise, we need to take partial derivative of the above equation with respect to m and b as these are the only two variables. That is

$$\frac{\partial SE}{\partial m} = 0 \; and \; \frac{\partial SE}{\partial b} = 0$$

$$\frac{\partial SE}{\partial m} = -2n\overline{xy} + 2n\overline{x^2}m + 2bn\overline{x} = 0$$

$$\frac{\partial SE}{\partial b} = -2n\overline{y} + 2mn\overline{x} + 2nb = 0$$

The equations on further simplification look similar to given two equations with two unknowns kind of problem.

$$\overline{x^2}m + b\overline{x} = \overline{xy}$$
$$m\overline{x} + b = -\overline{y}$$

We can see from the above equations that for optimal m and b, we are going to get a $\overline{x}$ and $\overline{y}$

$$m = \frac{\overline{xy} - \overline{x}\overline{y}}{\overline{x^2} - \overline{x}^2}$$
$$b = \overline{y} - m\overline{x}$$

7.2.2 Scalding Implementation

Using scalding to implement this is straight forward, We use the Scalding Fields APIs for its convenience.

Steps for solving a Univariate Linear Regression problem:

Step 1: Initialize a Scalding Job, **LRJob** which extends the Job class, that implements the necessary I/O and the Flow Definition for the execution of the job.

Note: It is important to note that every Scalding Job must ensure that there is Source, and Sink defined for successful execution.

```
import com.twitter.scalding._
class LRJob(args: Args) extends Job(args) {

// Code Goes Here

}
```

Step 2: We need define the *Source* for the input and read the data, as shown in the table 7.11, The dataset is stored in in.csv, Look at the github repository for this book.

The read method creates a Pipe, refer Cascading [29]

```
import com.twitter.scalding._
class LRJob(args: Args) extends Job(args) {
val input = Csv("in.csv", ",",   ('x, 'y))
    .read
    .write(Tsv("sampleinout.tsv"))
}
```

Step 3: In Section 7.2.1, we have derived and recognized certain variables for Slope and Intercept calculations. One such variable is the product of (x,y), over which we need to take the mean.

This code uses the Scalding Fields API to map through each row of the dataset and create a new Field with the product of *x* and *y*.

```
import com.twitter.scalding._
class LRJob(args: Args) extends Job(args) {
val input = Csv("in.csv", ",",   ('x, 'y))
    .read
    .map(('x, 'y) -> 'xy){
      f: (Double, Double) =>
      val (x,y) = f
      (x * y)
    }
    .write(Tsv("out.tsv"))
}
```

Sample Output: The following columns are represented by Fields ' x, ' y, and
' xy

Table 7.3: Product X and y

X	y	X * y
6.1101	17.592	107.4888792
5.5277	9.1302	50.46900654
8.5186	13.662	116.3811132
7.0032	11.854	83.015932799
5.8598	6.8233	39.98317334
8.3829	11.886	99.639149398
7.4764	4.3483	32.50963012
8.5781	12	102.937199
6.4862	6.5987	42.8004879404
5.0546	3.8166	19.29138636

Step 4: In this step, we determine the $\bar{x}$, read as Mean(x). To achieve this, we either
perform the sum of all the x and divide by the count traditionally, or use the existing
Scalding API to calculate the mean.

Scalding provides Grouping API, called _.sizeAveStdev that takes as input
a field of type Fields $\longrightarrow$ Count, Mean, and Standard Deviation(σ) as the output.

We perform a groupAll operation, This creates a single group consisting of the entire
pipe. Think three times before using this function on Hadoop. This removes the ability to do any parallelism in the reducers. That said, accumulating a global variable
may require it; in this case, we need to create a single variable in order to calculate
the mean and standard deviation.

The Api looks like below:

```
pipe.groupAll{ group => ... }
```

Example of such an operation. To find the total number of words in the vocabu-
lory.

```
val vocabSize = wordCounts.groupAll { _.size }
```

We also know that $\sigma^2 = variance$, we also calculate this by performing a simple
`math.pow(xsigma,2)`, using the scala libraries. Variance with respect to the
variable x can be written as, (Square of the mean - mean of the Squares) which we
have seen as the denominator of slope m.

We then project the fields `xmean` and `xvariance` out to the file *MVX.tsv*

$$variance(x) = \overline{(x^2)} - \overline{(x)}^2$$

```
import com.twitter.scalding._
class LRJob(args: Args) extends Job(args) {

val input = Csv("in.csv", ",",  ('x, 'y))
    .read
    .map(('x, 'y) -> 'xy){
      f: (Double, Double) =>
      val (x,y) = f
      (x * y)
    }

val MVofx = input
    .groupAll{
    _.sizeAveStdev('x -> ('count, 'xmean, 'xsigma))
    }
    .map('xsigma -> 'xvariance){
      (xsigma :Double) => (math.pow(xsigma,2))
    }
    .project('xmean, 'xvariance)
    .write(Tsv("MVX.tsv"))
}
```

Sample Output: The columns are represented internally in Scalding as fields
`xmean` and `xvariance`

Table 7.4

Mean(x)	Variance(x)
8.159800006	14.821606782061853

Step 5: As discussed in the previous step, we use the `sizeAveStdev` API to find
the `Count`, `Mean`, and `StdDeviation` of y. We then repeat the same steps as in
Step 4, but for the variable y.

We project the field `ymean` out to *MVY.tsv*

```
import com.twitter.scalding._
class LRJob(args: Args) extends Job(args) {
```

```
val input = Csv("in.csv", ",",   ('x, 'y))
    .read
    .map(('x, 'y) -> 'xy){
     f: (Double, Double) =>
     val (x,y) = f
     (x * y)
    }

val MVofx = input
    .groupAll{
     _.sizeAveStdev('x -> ('count, 'xmean, 'xsigma))
    }
    .map('xsigma -> 'xvariance){
      (xsigma :Double) => (math.pow(xsigma,2))
    }
    .project('xmean, 'xvariance)

val MVofy = input
    .groupAll{
     _.sizeAveStdev('y -> ('count, 'ymean, 'ysigma))
    }
    .project('ymean)
    .write(Tsv("MVY.tsv"))
}
```

Sample Output: The columns internally in Scalding are represented as the Field
ymean

Table 7.5: Mean of Field y

Mean(y)
5.839135051546394

Step 6: We repeat Step 5 on the field xy, which was calcluated in the Step 3, to find
the mean of the field xy and project it to file *MVXY.tsv*

```
import com.twitter.scalding._
class LRJob(args: Args) extends Job(args) {

val input = Csv("in.csv", ",",   ('x, 'y))
    .read
    .map(('x, 'y) -> 'xy){
     f: (Double, Double) =>
     val (x,y) = f
     (x * y)
    }

val MVofx = input
    .groupAll{
```

```
    _.sizeAveStdev('x -> ('count, 'xmean, 'xsigma))
  }
  .map('xsigma -> 'xvariance){
    (xsigma :Double) => (math.pow(xsigma,2))
  }
  .project('xmean, 'xvariance)

val MVofy = input
  .groupAll{
    _.sizeAveStdev('y -> ('count, 'ymean, 'ysigma))
  }
  .project('ymean)

val MVofxy = input
  .groupAll{
      _.sizeAveStdev('xy -> ('count, 'xymean, 'xysigma))
  }
  .project('xymean)
  .write(Tsv("MVXY.tsv"))
}
```

Sample Output: The column internally in Scalding is represented as Field `Mean (xy)`

Table 7.6: Mean of Field xy

Mean(xy)
65.3288497455567

Step 7: Until now, we have calculated $Mean(x) : \bar{x}, Variance(x) : \sigma^2(x), Mean(y) : \bar{y}$ and $Mean(xy) : \overline{xy}$.

Now we need to combine them into one Pipe, that is, we achieve this by performing a cross product of *MVofx*, *MVofy*, and *MVofxy*. Scalding provides a convenient API as below:

```
pipe1.crossWithTiny(pipe2)
```

Since `crossWithTiny` takes a source and a sink, Think of it as connecting two pipes as in Linux. We need to perform this twice. Now we perform two map operations:

- Find Slope m and identify it as Field `slope`

$$m = \frac{Mean(xy) - (Mean(x) * Mean(y))}{Variance(x)}$$

- Find Intercept b and identify it as Field `intercept`

$$b = Mean(y) - (m * Mean(x))$$

Project the `slope` and `intecept` out to Final.tsv

```
import com.twitter.scalding._
class LRJob(args: Args) extends Job(args) {

val Input = Csv("in.csv", ",",   ('x, 'y))
    .read
    .map(('x, 'y) -> 'xy){
      f: (Double, Double) =>
      val (x,y) = f
      (x * y)
    }

val MVofx = Input
    .groupAll{
    _.sizeAveStdev('x -> ('count, 'xmean, 'xsigma))
    }
    .map('xsigma -> 'xvariance){
      (xsigma :Double) => (math.pow(xsigma,2))
    }
    .project('xmean, 'xvariance)

val MVofy = Input
    .groupAll{
      _.sizeAveStdev('y -> ('count, 'ymean, 'ysigma))
    }
    .project('ymean)

val MVofxy = Input
    .groupAll{
      _.sizeAveStdev('xy -> ('count, 'xymean, 'xysigma))
    }
    .project('xymean)
}
val Final = MVofx
    .crossWithTiny(MVofy)
    .crossWithTiny(MVofxy)
    .map(('xmean, 'xvariance, 'ymean, 'xymean)->'slope){
      f: (Double, Double, Double, Double) =>
      val (mx, vx, my, mxy) = f
      ( (mxy - ( mx * my )) / vx)
    }
    .map(('xmean, 'ymean, 'slope) -> 'intercept){
      f : (Double, Double, Double) =>
      val(mx, my, slope) = f
      (my - (slope * mx))
    }
    .project('slope, 'intercept)
    .write(Tsv("final.tsv"))
```

Sample Output: The columns are internally represented as `slope` and `intercept`.

Table 7.7: Slope and Intercept

Slope	Intercept
1.1930336441895903	-3.8957808783118333

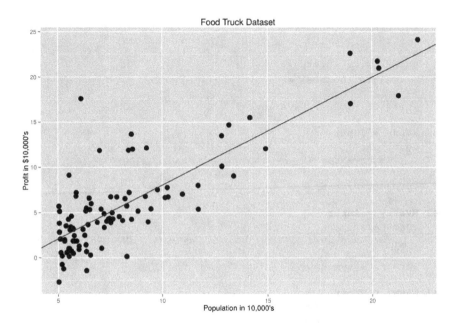

Fig. 7.3: Scalding: Regression Line for Food Truck Dataset

7.2.3 Spark Implementation

Steps for solving a Univariate Regression Problem:

Initialization of the Spark Project: In order for you to compile and execute Spark-based programs you need to create a Spark project. The steps involved are:

- Create a directory structure of `src/main/scala`

- Create a `project-name.sbt` file also called the scala build too file, that is used to declare project dependencies, track project versions, etc.

- Create the Scala Spark Class in the `scala` folder path created earlier.

- Use `sbt package` command to compile and build the project.

- Use the following code sample to submit the program to the Spark Cluster Master.

```
spark-submit --class <Class-Name> --master <cluster Url>
  <path/to/project.jar>
```

  - replace `<Class-Name>` with the Spark Class name you have used in the program

  - replace `cluster Url` with the path and port of the Spark Master, typically it is `spark://hostname:7077`.

  - `path/to/project.jar` is replaced typically by the jar target/scala-2.10/project.jar.

Step 1: Programming in Spark is very similar to programming in Regular Scala.

Here we create Spark Configuration, analogous to Scalding Job or Hadoop Configuration, it is used to set the various paraments as key-value pairs. You would create a new configuration using:

```
new SparkConf()
```

This loads default system properties and the classpath. You could also use

```
new SparkConf(loadDefaults: Boolean)
```

to skip loading external configuration settings. All the configuration methods in this Class support chaining, as in you can create a configuration like:

```
new SparkConf().setMaster("local").setAppName("My app")
```

Note: It is important to remember that once a Spark Conf. is created and passsed to the Spark Cluster, it is cloned and cannot be changed at runtime.

We also create an instance of `SparkContext`, with a defined configuration `conf`, which is a main entry point for untilizing the Spark Functionality. A `Spark Context` represents a connection to a Spark Clusters and exposes API's to create RDD(Resilient Distributed Datasets, refer Chapter 3), create Accumulators and Broadcast variables.

```
import org.apache.spark.SparkContext
import org.apache.spark.SparkContext._
import org.apache.spark.SparkConf

object SimpleApp {
  def main(args: Array[String]) {
```

```
val conf = new SparkConf().setAppName("Simple
                                      Application")
val sc = new SparkContext(conf)
}
```

Step 2: We read the input through the SparkContext instance `sc`, using a convenient method to read the text file. The API looks like this. It takes path to the file either in the Hadoop Distributed File System or the local system(available on all nodes) or any other Hadoop Supported file system. It returns an RDD of Strings.

```
textFile(path: String,
        minSplits: Int = defaultMinSplits): RDD[String]
```

The input, refer 7.11, is processed such that you map each row and split on ',', which returns an array of x's and y's, to seperate the x and y columns. Convert the String type of values to Double for Arithmentic operations following the process predefined.

The resulting RDD x, is then saved as a compressed TextFile using the `saveAsText File` method. The API definition looks like below, it takes the `path` to which the File is to saved and the compression `codec` for compression.

```
saveAsTextFile(path: String,
               codec: Class[_ <: CompressionCodec]):Unit

import org.apache.spark.SparkContext
import org.apache.spark.SparkContext._
import org.apache.spark.SparkConf

object SimpleApp {
  def main(args: Array[String]) {
    val conf = new SparkConf().setAppName("Simple
                                           Application")
    val sc = new SparkContext(conf)
    val input = sc.textFile("in.txt")
    val x = input
    .map( line => {
      val parts = line.split(',')
      (parts(0).toDouble)
    })
    x.saveAsTextFile("x.txt")
}
```

Sample Output : According to the code we have isolated the first column of the dataset, refer Table 7.11, as Independent variable x. as in Table 7.8

Step 3: Repeat Step 2, create a RDD of the y column, refer 7.11.

```
import org.apache.spark.SparkContext
import org.apache.spark.SparkContext._
import org.apache.spark.SparkConf
```

Table 7.8: RDD of x

X
6.1101
5.5277
8.5186
7.0032
5.8598
8.3829
7.4764
8.5781
6.4862
5.0546
5.7107
14.164

```
object SimpleApp {
  def main(args: Array[String]) {
    val conf = new SparkConf().setAppName("Simple
                                     Application")
    val sc = new SparkContext(conf)
    val input = sc.textFile("in.txt")
    val x = input
    .map( line => {
      val parts = line.split(',')
      (parts(0).toDouble)
    })

val y = input
    .map( line => {
      val parts = line.split(',')
      (parts(1).toDouble)
    })
    .saveAsTextFile("y.txt")
}
```

Sample Output: contents of *y.txt* looks like in Table 7.9

Step 4: Now that we have created and RDD's of x and y, we need the RDD of product of (x,y). The simplest way to do that is to map each row of the input dataset, refer 7.11, split each mapped row and multiply the resulting tuples.

Table 7.9: RDD of y

Y
17.592
9.1302
13.662
11.854
6.8233
11.886
4.3483
12
6.5987
3.8166
3.2522
15.505

```scala
import org.apache.spark.SparkContext
import org.apache.spark.SparkContext._
import org.apache.spark.SparkConf

object SimpleApp {
  def main(args: Array[String]) {
    val conf = new SparkConf().setAppName("Simple
                                          Application")
    val sc = new SparkContext(conf)
    val input = sc.textFile("in.txt")
    val x = input
    .map( line => {
      val parts = line.split(',')
      (parts(0).toDouble)
    })

val y = input
    .map( line => {
      val parts = line.split(',')
        (parts(1).toDouble)
    })
}

val xy = input
    .map( line => {
      val parts = line.split(',')
      (parts(0).toDouble * parts(1).toDouble)
    })
    .saveAsTextFile("xy.txt")
```

Table 7.10: Product X and y

X*y
107.4888792
50.46900654
116.3811132
83.015932799
39.98317334
99.639149398
32.50963012
102.937199
42.8004879404
19.29138636

Sample Output:

Step 5: In order to calculate the slope m, we need to calculate the *Mean(x)* : $\bar{x}$, *Variance(x)* : $\sigma^2(x)$, *Mean(y)* : $\bar{y}$, *and Mean(xy)* : $\overline{xy}$. Spark provides a convenient way to doing this without extra processing as in Scalding. Since the RDD's of x, y, and xy are of RDD[Double], the built-in Double RDD APIs provide convenient methods to find the mean, variance, and other results.

```
import org.apache.spark.SparkContext
import org.apache.spark.SparkContext._
import org.apache.spark.SparkConf

object SimpleApp {
  def main(args: Array[String]) {
    val conf = new SparkConf().setAppName("Simple
                                    Application")
    val sc = new SparkContext(conf)
    val input = sc.textFile("in.txt")
    val x = input
    .map( line => {
      val parts = line.split(',')
      (parts(0).toDouble)
    })

val y = input
    .map( line => {
      val parts = line.split(',')
        (parts(1).toDouble)
    })
}

val xy = input
    .map( line => {
      val parts = line.split(',')
        (parts(0).toDouble * parts(1).toDouble)
    })
```

```
val slope = (xy.mean - (x.mean * y.mean)) / x.variance

val intercept = y.mean - slope * x.mean

println (slope, intercept)
```

Output:

```
1.1930336441895903   -3.8957808783118333
```

Step 6: Plotting the graph, we use the R programming language assisted by the ggplot2 library.

```
#!/usr/bin/env Rscript
library(ggplot2)
x <- read.csv("in.txt", header=FALSE, sep=",")
names(x) <- c("x", "y")

ggplot() +
    geom_point(data=x, aes(x=x, y=y), size=4) +
    geom_abline(intercept=-3.89, slope=1.193, colour="blue") +
    ggtitle("Regression Line for Food Truck Dataset") +
    ggsave(file="LR.png", width=10, height=7)
```

The above code produces the following image:

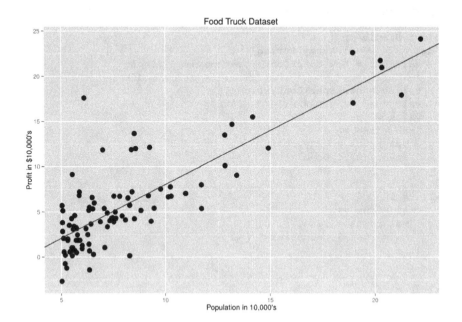

Fig. 7.4: Spark: Regression Line for Food Truck Dataset

7.2.4 Linear Regression: Gradient Descent Method

As we discussed earlier, Linear Regression is a process of fitting models to data. When the target variable we are trying to predict is continuous, we call it a **Regression** problem. When y can take on only a small number of discrete values we call it a **Classification** problem. Linear Regression assumes that the relationship between the X (independent variable) and y (dependent variable) is Linear, i.e., y = mX + b, where m is the slope and b is the intercept.

$$h_\theta(x) = \theta_0 + \theta_1 x$$

This can be represented as the following equation and usually called the Hypothesis. You can notice that expression is similar to the linear expression y = mx + b. The idea of Linear Regression is to find a combination of θ_0 and θ_1 such that h_θ is close to y for our training examples (x, y).

Formalizing this idea, is to solve a minimization problem over (θ_0, θ_1). We want the difference between ($h_\theta - y$) be small. One way we can do this is minimize the squared difference between ($h_\theta - y$)^2. We need to perform this operation to all the points in the dataset, we do this by the following equation

$$minimize(\theta_0, \theta_1) = \sum_{i=1}^{m} \left(h_\theta(x_i) - y_i \right)^2$$

A task to minimize the squared errors between the point and the line. This method is also popularly called as Ordinary Least Squares method(OLS).

We define a **Cost Function** ($J(\theta_0, \theta_1)$) such that:

$$J(\theta_0, \theta_1) = \sum_{i=1}^{m} \left(h_\theta(x_i) - y_i \right)^2$$

This Cost function is also called the Squared Error Cost Function. It is the most commonly used function as it works reasonably well for Linear Regression problems.

We use **Gradient Descent** to Minimize the Cost function, as you can observe in the discussion hence forth the description of the Gradient Descent is more generic and is used all over the place for various minimization problems. Gradient Descent starts with some initial θ, and repeatedly performs the following update:

$$\theta_j := \theta_j - \alpha \frac{\partial}{\partial \theta_j} J(\theta)$$

Note: This update is simulatneously performed for all values of $j = 0,1,2,3....,n$.

Here α is called the **Learning Rate**. This is a very intuitive algorithm that repeatedly takes a step toward the minimum of J. In order to implement this, we need to compute the partial derivative of $J(\theta)$. Let us compute if we had only one training example (x,y), so that we ignore the sum in definition of $J(\theta)$.

$$
\begin{aligned}
\frac{\partial}{\partial \theta_j} J(\theta_j) &= \frac{\partial}{\partial \theta_j} \frac{1}{2} (h_\theta(x) - y)^2 \\
&= 2 \cdot \frac{1}{2} (h_\theta(x) - y) \cdot \frac{\partial}{\partial \theta_j} (h_\theta(x) - y) \\
&= (h_\theta(x) - y) \cdot \frac{\partial}{\partial \theta_j} \left(\sum_{i=0}^{m} \theta_i x_i - y \right) \\
&= (h_\theta(x) - y) x_j
\end{aligned}
$$

For a single training example, this update gives [1]:

$$
\theta_j := \theta_j + \alpha (y - h_\theta(x)) x_j
$$

This is also called the Widrow-Hoff learning rule. This rule has several properties that seem natural and intuitive. For instance, the magnitude of the update is proportional to the error term $(y - h_\theta(x))$; for instance, if we encounter a training example for which the prediction nearly matches the value y, then we find there is little need to change the paramenters; in contrast, if we encounter a training example for which our prediction has a large error, then the parameters will need to make a large change.

We have derived the Least Mean Squares rule for when we had one training example. There are two ways to modify this to manage the entire training set. First replace it with the following:

Repeat until convergence {

$$
\theta_j := \theta_j + \alpha \sum_{i=1}^{m} (y - h_\theta(x)) x_j \quad (\forall j)
$$

}

You can verify the quantity in the summation in the update rule above is just $\partial(J(\theta))/\partial \theta_j$. So, this is simply a gradient descent on the original cost funtion J.

---

[1] We use the notation a := b to denote an operation (in a computer program) in which we set the value of a variable a to be equal to the value of b. In other words, this operation overwrites a with the value of b. In contrast, we will write a = b when we are asserting a statement of fact, that the value of a is equal to the value of b.

This method looks at every training example at every step, and is called **batch gradient descent**.

Note: Though gradient descent can have many local minima, the optimization problem we have defined for Linear Regression will have only one global minima; thus gradient descent always converges (assuming the learning rate α is not too large) to a global minimum.

For a simple example on how the gradient descent progresses, check the Figure 7.5

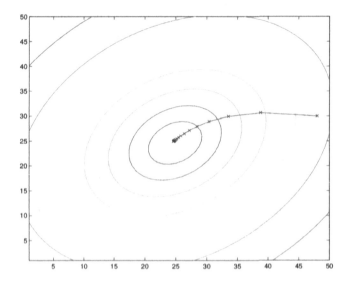

Fig. 7.5: Sample Gradient Descent moving towards a minimum

The ellipses shown in the Figure 7.5 are the contours of a quadratic function. Also shown is the trajectory taken by the gradient descent which was initialized at (48,30). The x's in the figure (joined by straight lines) mark the successive values of θ that gradient descent went through.

There is an alternative to the batch gradient descent method, consider the following algorithm

```
Loop {
        for i=1 to m{
```
$$\theta_j := \theta_j + \alpha\left(y - h_\theta(x)\right)x_j \qquad (\forall j)$$
```
        }
}
```

As per the above algorithm, we process each training set and update the parameters for each item according to the gradient error of that training example. This algorithm is called **Stochastic Gradient Descent** or incremental descent. The difference between this method and the previous Batch Descent is that in the later, before taking a single step the parameters are updated for all the training examples which turns out to be costly if m is large. Stochastic method makes progress right away with each training example it encounters. It is observed that in Stochastic Gradient Descent θ reaches the minimum must faster than in the Batch method. For these reasons, Stochastic method is preferred when the training set is large.

Note: Stochastic method may, however, never converge to the minimum, and the parameters in question, θ, may keep oscillating around the minimum $J(\theta)$, but in most scenarios the minimum obtained will be reasonably good approximations to the true minimum. [2]

7.2.5 Scalding Implementation

Map Reduce-based implementation of an iterative algorithm like the Linear Regression is not advisable because, the algorithm used to perform the minimization clearly indicates dependence; to be more clear, θ_{j+1} iteration depends on the value of θ_j which is not suitable for MapReduce paradigm. That said, there are ways in which you can do using Scalding.

- Chaining multiple Scalding Jobs

- Recursive Scalding Job

Both these methods are in no way efficient. We will concentrate on the first method for this implementation. Implementation is through the following steps:

- Input Preprocessing

Step 1 - Input Preprocessing: As we have seen before the input is of the format:

---

[2] While it is more common to run stochastic gradient descent as we have described it and with a fixed learning rate α, by slowly letting the learning rate decrease to zero as the algorithm runs, it is also possible to ensure that the parameters will converge to the global minimum rather then merely oscillate around the minimum.

Table 7.11: Sample Dataset

Population (10,000's)	Profit ($10,000's)
6.1101	17.592
5.5277	9.1302
8.5186	13.662
7.0032	11.854
5.8598	6.8233
8.3829	11.886
7.4764	4.3483
8.5781	12
6.4862	6.5987
5.0546	3.8166

In order for us to implement Linear Regression using chained Scalding Jobs using Batch Gradient Method, we need to initialize the starting point of the parameters θ_0 and θ_1, where θ_0 is the *intercept* and θ_1 is the slope. The hypothesis is:

$$y = \theta_0 + \theta_1 * x$$

We initialize the $\theta_0 = 0.0$ and $\theta_1 = 0.0$. A clever way to overcome the dependency issue mentioned previously is we add the $\theta's$ as additional columns in the input. A simple way to do this is by creating a Scalding Job that:

- Reads the input dataset

- For each line of input, for each (x,y) add θ_0 and θ_1 as extra fields

- Write the result to file *swapin.tsv*

```
import com.twitter.scalding._
class LRInputJob(args: Args) extends Job(args) {

val input = Csv("in.csv", ",", ('x, 'y))
    .read

val init = input
    .map(('x, 'y) -> ('t0, 't1)){
      f : (Double, Double) =>
      val (x, y) = f
      (0.0, 0.0)
    }

    .write(Tsv("swapin.tsv"))
```

Sample Output: This is how the input looks like after pre-processing as shown in Table 7.12

Table 7.12

x	y	θ_0	θ_1
6.1101	17.592	0.0	0.0
5.5277	9.1302	0.0	0.0
8.5186	13.662	0.0	0.0
7.0032	11.854	0.0	0.0
5.8598	6.8233	0.0	0.0
8.3829	11.886	0.0	0.0
7.4764	4.3483	0.0	0.0
8.5781	12	0.0	0.0
6.4862	6.5987	0.0	0.0
5.0546	3.8166	0.0	0.0

Step 2: Now that the initial θ values have been set we need to devise a procedure such that the output of first Scalding Job is the input to the Second Job and so on... This means we need to a mechanism to swap the output and the input files. Carefully looking at the source code of the `Job` class we find two functions that can enable us to execute this:

- `clone` method.

   The method definition looks like this, it is seen that the clone method takes the arguments used to launch a new Job.

   ```
   def clone(nextargs: Args): Job
   ```

- `next` method.

   The method definition looks like this, We implement this method in order to signal Scalding to run some other Job (defined in this method) after the current Job. Ideally you use it for operations like a clean-up. This `next` job will not be executed until the successful completion of the current Job, safe to say you can easily debug in the event of any errors.

   ```
   def next: Option[Job] = None
   ```

To implement the swapfiles mechanism we use a temporary file that will act as a buffer between Jobs. The reason we need a tempFile is that a `Source`(in this case inputFile) cannot be `Sink`(outputFile). The way it works is:

$$inputFile = tempFile$$
$$tempFile = outputFile$$
$$outputFile = inputFile$$

The Scalding code snipped looks like this:

```
override def next: Option[Job] = {
    val nextArgs = args +
      ("input", Some(args("output"))) +
      ("temp", Some(args("output"))) +
      ("output", Some(args("temp")))

    Some(clone(nextArgs))
}
```

The above code, though cannot exist independently, carefully explains how you can use the clone method and next method together to create multiple jobs. The above code swaps the files once. But Linear Regression includes a set iterations that is needed to arrive at a minimum. To allow for this we add another command line argument called JobCount. Initially we set this number to MAX_ITERATIONS, as you clone the next Job you decrease the JobCount by 1 and have exit case for when the JobCount is 0.

```
import com.twitter.scalding._
class SwapJob(args: Args) extends Job(args) {

val JOB_COUNT = args("jobCount").toInt

override def next: Option[Job] = {
    val nextArgs = args +
    ("input", Some(args("output"))) +
    ("temp", Some(args("output"))) +
    ("output", Some(args("temp"))) +
    ("jobCount", Some((JOB_COUNT - 1).toString))

    if ((JOB_COUNT > 1)) {
      Some(clone(nextArgs))
    }
    else {
      None
    }
}

    val input = Tsv(args("input"))
          .read
          .write(Tsv(args("output")))
  }
```

Ouput: As you can observe the Job Count set to 2 creates two Jobs, with the files swapped.

```
\$ ../scripts/scald.rb --local swapfiles.scala --input swapin.tsv \
    --output swapout.tsv --temp swaptemp.tsv --jobCount 2

INFO property.AppProps: using app.id: B7E22898CDF7439AAC94757FE477B0F0
INFO util.Version: Concurrent, Inc - Cascading 2.5.4
... starting
...   source: FileTap["TextDelimited[[UNKNOWN]->[ALL]]"]["swapin.tsv"]
```

```
...   sink: FileTap["TextDelimited[[UNKNOWN]->[ALL]]"]["swapout.tsv"]
...   parallel execution is enabled: true
...   starting jobs: 1
...   allocating threads: 1
... starting step: local
... starting
...   source: FileTap["TextDelimited[[UNKNOWN]->[ALL]]"]["swapout.tsv"]
...   sink: FileTap["TextDelimited[[UNKNOWN]->[ALL]]"]["swaptemp.tsv"]
...   parallel execution is enabled: true
...   starting jobs: 1
...   allocating threads: 1
... starting step: local
```

Step 3: We initialize a Scalding that calculates slope and intercept through batch gradient descent method. First we initialize a Scalding Job that extends the Job class.

```scala
import com.twitter.scalding._
class LRBGDJob(args: Args) extends Job(args) {

// Code Goes Here

}
```

Step 4: We hook the previously explained `clone` and `next` Job methods implementations.

```scala
import com.twitter.scalding._
class LRBGDJob(args: Args) extends Job(args) {
  val JOB_COUNT = args("jobCount").toInt

  override def next: Option[Job] = {
    val nextArgs = args + ("input", Some(args("output"))) +
    ("temp", Some(args("output"))) +
    ("output", Some(args("temp"))) +
    ("jobCount", Some((JOB_COUNT - 1).toString))

    if ((JOB_COUNT > 1)) {
      Some(clone(nextArgs))
    }
    else {
      None
    }
  }
}
```

Step 5: Read the input and initialize the Fields x, y, θ_0 and θ_1

```scala
import com.twitter.scalding._
class LRBGDJob(args: Args) extends Job(args) {
  val JOB_COUNT = args("jobCount").toInt

  override def next: Option[Job] = {
    val nextArgs = args + ("input", Some(args("output"))) +
```

```
      ("temp", Some(args("output"))) +
      ("output", Some(args("temp"))) +
      ("jobCount", Some((JOB_COUNT - 1).toString))

    if ((JOB_COUNT > 1)) {
      Some(clone(nextArgs))
    }
    else {
      None
    }
  }

  val input = Tsv(args("input"), ('x, 'y, 't0, 't1))
      .read
      .write(Tsv("Testout.tsv"))
}
```

Step 6: Calculate the predictions and the errors for each pair of (x,y)

$$prediction = \theta_0 + \theta_1 * x$$
$$error(\theta_0) = prediction - y$$
$$error(\theta_1) = (prediction - y) * x$$

The code looks like this:

```
import com.twitter.scalding._
class LRBGDJob(args: Args) extends Job(args) {
  val JOB_COUNT = args("jobCount").toInt

  override def next: Option[Job] = {
    val nextArgs = args + ("input", Some(args("output"))) +
    ("temp", Some(args("output"))) +
    ("output", Some(args("temp"))) +
    ("jobCount", Some((JOB_COUNT - 1).toString))

    if ((JOB_COUNT > 1)) {
      Some(clone(nextArgs))
    }
    else {
      None
    }
  }

  val input = Tsv(args("input"), ('x, 'y, 't0, 't1))
      .read

  val errors = input
      .map(('x, 'y, 't0, 't1) -> ('e1, 'e2)){
        f : (Double, Double, Double, Double) =>
```

```
      val (x, y, t0, t1) = f
      val prediction = (t0 + (t1 * x))
      val e1 = prediction - y
      val e2 = (prediction - y) * x
      (e1, e2)
    }
    .write(Tsv("lrerrors.tsv"))
}
```

Output: Shows the Error(θ_0) and Error(θ_1)

Table 7.13: Output with θ errors

x	y	θ_0	θ_1	error θ_0	error θ_1
6.1101	17.592	0.0	0.0	-17.592	-107.4888792
5.5277	9.1302	0.0	0.0	-9.1302	-50.46900654
8.5186	13.662	0.0	0.0	-13.662	-116.3811132
7.0032	11.854	0.0	0.0	-11.854	-83.015932799
5.8598	6.8233	0.0	0.0	-6.8233	-39.98317334
8.3829	11.886	0.0	0.0	-11.886	-99.639149398
7.4764	4.3483	0.0	0.0	-4.3483	-32.50963012
8.5781	12	0.0	0.0	-12.0	-102.9371999
6.4862	6.5987	0.0	0.0	-6.5987	-42.80044004
5.0546	3.8166	0.0	0.0	-3.8166	-19.29138636

Step 6: Calculate the sums of the errors as in

$$sum(error\theta_0) = \sum_{i=1}^{m} (y - h_\theta(x))$$

$$sum(error\theta_1) = \sum_{i=1}^{m} (y - h_\theta(x))x_j$$

The code looks like below; we perform a `groupAll` and a `sum` operation on the field `'e1` and `'e2` respectively.

```
import com.twitter.scalding._
class LRBGDJob(args: Args) extends Job(args) {
  val JOB_COUNT = args("jobCount").toInt

  override def next: Option[Job] = {
    val nextArgs = args + ("input", Some(args("output"))) +
    ("temp", Some(args("output"))) +
    ("output", Some(args("temp"))) +
    ("jobCount", Some((JOB_COUNT - 1).toString))
```

```
      if ((JOB_COUNT > 1)) {
        Some(clone(nextArgs))
      }
      else {
        None
      }
   }

  val input = Tsv(args("input"), ('x, 'y, 't0, 't1))
      .read

  val errors = input
      .map(('x, 'y, 't0, 't1) -> ('e1, 'e2)){
        f : (Double, Double, Double, Double) =>
        val (x, y, t0, t1) = f
        val prediction = (t0 + (t1 * x))
        val e1 = prediction - y
        val e2 = (prediction - y) * x
        (e1, e2)
      }

  val sums = errors
      .groupAll{
        _.sum[Double]('e1 -> 'se1)
        .sum[Double]('e2 -> 'se2)
      }
      .write(Tsv("sums.tsv"))
}
```

Sample Output: This is the output for the first iteration:

Table 7.14

sum(e1)	sum(e2)
-566.3960999999998	-6336.898425319003

Step 7: We need to make this summed values available to perform the θ_0 and θ_1 updates.

To achieve this we use Scaldings crossWithTiny that performs the cross product between two Pipes.

```
import com.twitter.scalding._
class LRBGDJob(args: Args) extends Job(args) {
  val JOB_COUNT = args("jobCount").toInt

  override def next: Option[Job] = {
    val nextArgs = args + ("input", Some(args("output"))) +
      ("temp", Some(args("output"))) +
      ("output", Some(args("temp"))) +
```

```
    ("jobCount", Some((JOB_COUNT - 1).toString))

    if ((JOB_COUNT > 1)) {
      Some(clone(nextArgs))
    }
    else {
      None
    }
}

val input = Tsv(args("input"), ('x, 'y, 't0, 't1))
    .read

val errors = input
    .map(('x, 'y, 't0, 't1) -> ('e1, 'e2)){
       f : (Double, Double, Double, Double) =>
       val (x, y, t0, t1) = f
       val prediction = (t0 + (t1 * x))
       val e1 = prediction - y
       val e2 = (prediction - y) * x
       (e1, e2)
    }

val sums = errors
    .groupAll{
       _.sum[Double]('e1 -> 'se1)
       .sum[Double]('e2 -> 'se2)
    }
val errorswithsum = errors
    .crossWithTiny(sums)
    .write(Tsv("errorswithsum.tsv"))
}
```

Sample Output: Added new Fields ' se1 and ' se2 to the outputFile.

Table 7.15: Output with sums of θ errors

x	y	θ_0	θ_1	error θ_0	error θ_1	sum error θ_0	sum of error θ_1
6.1101	17.592	0.0	0.0	-17.592	-107.4888792	-566.39609998	-6336.8984253
5.5277	9.1302	0.0	0.0	-9.1302	-50.46900654	-566.39609998	-6336.8984253
8.5186	13.662	0.0	0.0	-13.662	-116.3811132	-566.39609998	-6336.8984253
7.0032	11.854	0.0	0.0	-11.854	-83.015932799	-566.3960998	-6336.8984253
5.8598	6.8233	0.0	0.0	-6.8233	-39.98317334	-566.39609998	-6336.8984253
8.3829	11.886	0.0	0.0	-11.886	-99.639149398	-566.3960998	-6336.8984253
7.4764	4.3483	0.0	0.0	-4.3483	-32.50963012	-566.39609998	-6336.8984253
8.5781	12	0.0	0.0	-12.0	-102.9371999	-566.39609998	-6336.8984253
6.4862	6.5987	0.0	0.0	-6.5987	-42.80044004	-566.39609999	-6336.8984253
5.0546	3.8166	0.0	0.0	-3.8166	-19.29138636	-566.39609999	-6336.8984253

Step 8: Update the θ_0 and θ_1 as in:

$$\theta_0 = \theta_0 - \alpha \cdot \frac{1}{m} \cdot \sum_{i=1}^{m} (y - h_\theta(x))$$

$$\theta_1 = \theta_1 - \alpha \cdot \frac{1}{m} \cdot \sum_{i=1}^{m} (y - h_\theta(x)) \cdot x$$

```scala
import com.twitter.scalding._
class LRBGDJob(args: Args) extends Job(args) {
  val JOB_COUNT = args("jobCount").toInt

  override def next: Option[Job] = {
    val nextArgs = args + ("input", Some(args("output"))) +
      ("temp", Some(args("output"))) +
      ("output", Some(args("temp"))) +
      ("jobCount", Some((JOB_COUNT - 1).toString))

    if ((JOB_COUNT > 1)) {
      Some(clone(nextArgs))
    }
    else {
      None
    }
  }

  val input = Tsv(args("input"), ('x, 'y, 't0, 't1))
      .read

  val errors = input
      .map(('x, 'y, 't0, 't1) -> ('e1, 'e2)){
        f : (Double, Double, Double, Double) =>
        val (x, y, t0, t1) = f
        val prediction = (t0 + (t1 * x))
        val e1 = prediction - y
        val e2 = (prediction - y) * x
        (e1, e2)
      }

  val sums = errors
      .groupAll{
        _.sum[Double]('e1 -> 'se1)
        .sum[Double]('e2 -> 'se2)
      }
  val errorswithsum = errors
      .crossWithTiny(sums)

}
  val newthetas = errorswithsum
      .map(('t0, 't1, 'se1, 'se2) -> ('nt0, 'nt1)){
        f : (Double, Double, Double, Double) =>
```

```
    val(t0, t1, se1, se2) = f
    val nt0 = t0 - (0.01) * (1.0 / 97) * se1
    val nt1 = t1 - (0.01) * (1.0 / 97) * se2
    (nt0, nt1)
  }
  .project('x, 'y, 'nt0, 'nt1)
  .write(Tsv(args("output")))
```

Output : Looking at the output file we see the columns θ_0 and θ_1 to find values:

```
\$ ../scripts/scald.rb --local LRBGDJob.scala --input swapin.tsv
   --output swapout.tsv --temp swaptemp.tsv --jobCount 4250

-3.8957808783118333 1.1930336441895903
```

7.2.6 Spark Implementation

Spark is perfectly suited for applications that need to be distributed and perform iterations like in Linear Regression where there is dependency.

Step 1: Initialize the SparkContext and Spark related Configuration. SparkContext acts as an entry point in to the spark engine. It exposes several constrcuts and methods to ease creation, manipulation of large data in-memory. Programming in Spark is very similar to Scala programming, we define an Object and a main function to begin execution.

```
import org.apache.spark.SparkContext
import org.apache.spark.SparkContext._
import org.apache.spark.SparkConf

object BGD {
    def main(args: Array[String]) {
      val conf = new SparkConf()
            .setAppName("Simple Application")
      val sc = new SparkContext(conf)
    }
}
```

Step 2: Read the input using SparkContext's textFile method. The API for this command takes the path of the inputFile and returns an RDD(Resilient Distributed Dataset) on which you can perform actions.

```
textFile(path: String,
        minSplits: Int = defaultMinSplits): RDD[String]

import org.apache.spark.SparkContext
import org.apache.spark.SparkContext._
```

```
import org.apache.spark.SparkConf

object BGD {
  def main(args: Array[String]) {
    val conf = new SparkConf()
          .setAppName("Simple Application")
    val sc = new SparkContext(conf)
    val input = sc.textFile("in.txt")
  }
}
```

Step 3: Pre-process the input so that it is available for subsequent computation. The input is mapped such that each row in the inputFile is split on the ',' and resulting Array is stored as tuples.

```
object BGD {
  def main(args: Array[String]) {
    val conf = new SparkConf()
          .setAppName("Simple Application")
    val sc = new SparkContext(conf)
    val input = sc.textFile("in.txt")

    val xy = input
          .map( line => {
              val parts = line.split(',')
              (parts(0).toDouble, parts(1).toDouble)
          })
  }
}
```

Sample Output: Shows the Array of Tuples of (x,y).

```
 WARN NativeCodeLoader: Unable to load native-hadoop
library for your platform...
WARN LoadSnappy: Snappy native library not loaded
 INFO FileInputFormat: Total input paths to process : 1
INFO SparkContext: Starting job: collect at <console>:17
.
.
.
INFO DAGScheduler: Stage 0 (collect at <console>:17)
finished in 0.159 s
INFO SparkContext: Job finished:
collect at <console>:17, took 0.275020439 s
res0: Array[(Double, Double)] = Array((6.1101,17.592),
 (5.5277,9.1302), (8.5186,13.662), (7.0032,11.854),
 (5.8598,6.8233), (8.3829,11.886), (7.4764,4.3483),
 (8.5781,12.0), (6.4862,6.5987), (5.0546,3.8166),
 (5.7107,3.2522), (14.164,15.505), (5.734,3.1551),
 (8.4084,7.2258), (5.6407,0.71618), (5.3794,3.5129),
 (6.3654,5.3048), (5.1301,0.56077),
 (6.4296,3.6518), (7.0708,5.3893), (6.1891,3.1386),
 (20.27,21.767), (5.4901,4.263), (6.3261,5.1875),
```

```
(5.5649,3.0825), (18.945,22.638), (12.828,13.501),
(10.957,7.0467), (13.176,14.692), (22.203,24.147),
(5.2524,-1.22), (6.5894,5.9966), (9.2482,12.134),
(5.8918,1.8495), (8.2111,6.5426), (7.9334,4.5623), . . .
```

Step 4: Set the initial θ_0 and θ_1 values to 0.0, also find the size of the input

```scala
object BGD {
  def main(args: Array[String]) {
    val conf = new SparkConf()
         .setAppName("Simple Application")
    val sc = new SparkContext(conf)
    val input = sc.textFile("in.txt")

    val xy = input
         .map( line => {
             val parts = line.split(',')
             (parts(0).toDouble, parts(1).toDouble)
         })

    var m = input.count().toDouble

    var theta0 : Double = 0.0
    var theta1 : Double = 0.0

    var e1sum : Double = 0.0
    var e2sum : Double = 0.0
  }
}
```

Step 5: Since Spark is an in-memory processing platform, it is inviting to us to write a simple for loop to perform the regression. Updating θ values happens inside this loop.

$$sum(error\theta_0) = \sum_{i=1}^{m} (y - h_\theta(x))$$

$$sum(error\theta_1) = \sum_{i=1}^{m} (y - h_\theta(x))x_j$$

We implement the $sum(error\theta_0)$ using the following code snippet:

```scala
e1sum = xy.map( line => {
      var prediction s= theta0 + theta1 * line._1
      var diff = prediction - line._2
      (diff)
   }).sum()
```

We implement the $sum(error\theta_1)$ using the following code snippet:

```
e2sum = xy.map( line => {
    var prediction = theta0 + theta1 * line._1
    var diff = (prediction - line._2) * line._1
    (diff)
}).sum()
```

Updating the $\theta's$:

```
theta0 = theta0 - 0.01 * ( 1.0 / m ) * e1sum
theta1 = theta1 - 0.01 * ( 1.0 / m ) * e2sum
```

Plugging all this code into a for loop iterating over the chosen number inorder for the $\theta's$ to converge.

```
object BGD {
  def main(args: Array[String]) {
    val conf = new SparkConf()
        .setAppName("Simple Application")
    val sc = new SparkContext(conf)
    val input = sc.textFile("in.txt")

    val xy = input
        .map( line => {
            val parts = line.split(',')
            (parts(0).toDouble, parts(1).toDouble)
        })

    var m = input.count().toDouble

    var theta0 : Double = 0.0
    var theta1 : Double = 0.0

    var e1sum : Double = 0.0
    var e2sum : Double = 0.0

    for( i <- 1 to 4250){
      e1sum = xy.map(
        line => {
          var prediction = theta0 + theta1 * line._1
          var diff = prediction - line._2
          (diff)
        }).sum()

      e2sum = xy.map(
        line => {
          var prediction = theta0 + theta1 * line._1
          var diff = (prediction - line._2) * line._1
          (diff)
        }).sum()
      theta0 = theta0 - 0.01 * ( 1.0 / m ) * e1sum
      theta1 = theta1 - 0.01 * ( 1.0 / m ) * e2sum
      }
    println(theta0, theta1)
  }
}
```

Output: Executing the following command will run the above code in the Spark engine:

```
sbt package
../bin/spark-submit --class "BGD" --master <path/to/spark/master> \
target/scala-2.10/bgd.jar
```

-3.8957808783118333 1.1930336441895903

Plotting the graph using the R code mentioned earlier in the chapter:

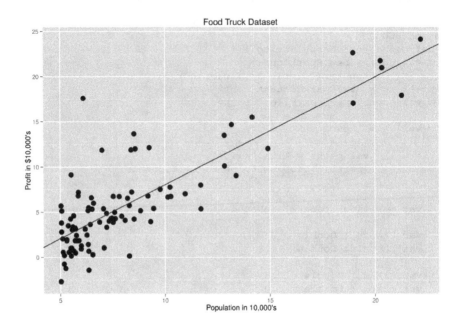

Fig. 7.6: Spark: Regression Line for Food Truck Dataset using Batch Gradient Descent

Problems

7.1. Use the dataset in the *problemset.txt* in the Linear Regression section on this book's github repository to determine the following:
(a) Find the relationship between *University GPA* and *High School GPA*
(b) Find the relationship between *Math SAT Score*, *Verbal SAT Score* and *University GPA*

7.2. This chapter implements Batch Gradient Descent, but for obvious reasons it is not scalable. A scalable version is presented in theory in the chapter called Stochastic Gradient Descent(SGD). Your task is to modify the code presented here, both in Scalding and Spark to implement the SGD algorithm.

References

1. Armstrong, J. Scott (2012). "Illusions in Regression Analysis". International Journal of Fore-casting (forthcoming) 28 (3): 689.
2. *Cascading Pipes* http://docs.cascading.org/cascading/1.2/javadoc/cascading/pipe/Pipe.html
3. Bartlett, G., Stewart, J. D., Tamblyn, R., & Abrahamowicz, M. (1998). *Normal distributions of thermal and vibration sensory thresholds.* Muscle & nerve, 21(3), 367-374.
4. Rencher, Alvin C., and William F. Christensen. *Methods of multivariate analysis.* Vol. 709. John Wiley & Sons, 2012.

Chapter 8

Case Study IV: Recommender System using Scalding and Spark

Recommender Systems are software tools that are used to suggest items of use to users based on certain assumptions [1] [2] . The *item* here refers to an entity that the system recommends to the users, and accordingly the recommender system's design, GUI, recommendation technique are dependent on the specific type of *item* in the discussion.

Recommender systems are directed at people who typically lack the experience to differentiate among the many alternatives of items found on the web. For example, in a popular site for selling books, like Amazon, a recommender system is used to suggest books to individual users. The suggestions are diverse and are personalized to individual users [3]. There are simpler recommendations that are not personalized like the top 10 selections of books or CDs. Although these recommendations are useful, they are not important in Recommender systems research.

8.1 Recommender Systems

Recommendations are usually represented as ranked lists of items. The ranking of these items are based on user's preferences and other constraints. In computing this ranking, the system captures the user's preferences either explicitly by ratings for products, or implicitly by user's actions, navigation to a product page as a sign of preference.

It is a common practice to rely on the recommendations provided by others in making a purchase or using a resource . Recommender Systems arise out of this practice [4] [5]. Employees rely on recommendation letters while hiring, it is common to rely on peer's suggestion on what to read, movie reviews are important in choosing a movie, etc. Collaborative-filtering is a strategy used in recommender systems that depends on the suggested item from a group of people with similar tastes as the cur-

© Springer International Publishing Switzerland 2015
K.G. Srinivasa and A.K. Muppalla, *Guide to High Performance
Distributed Computing*, Computer Communications and Networks,
DOI 10.1007/978-3-319-13497-0_8

rent active user and if the user accepts this suggestion, any future suggestion from these group of people is relevant. Recommender Systems mimic this behavior by applying algorithms that leverage recommendations from a community of users to an active user.

Due to explosive growth of e-commerce businesses on the Internet, the users are overwhelmed by the choices they provide leading them to make poor decisions [6]. Recommender systems are used in coping with this information over load problem. A recommender system can address this problem by pointing the user to not-yet-experienced item that may be relevant to the users current task. Recommender systems use various criteria in making suggestions like the users context, need, some data can be captured by the implicit user feedback. This data is stored in the recommender systems database and used for further the recommendation engine and suggest items.

8.1.1 Objectives

It is important to refine the definition given above to deeply understand the range of possible roles that a recommender system can play. We need to differentiate between a recommender system's role and user of the system. For example, a travel intermediary like Expedia.com uses recommender systems to increase the sale of the hotel rooms, to increase the number of tourists to a destination, while the users want to find a suitable hotel and interesting attractions in a destination upon using the recommender system [7]. There are various reason why a service provider may want to exploit this systems:

1. Increase the number of items sold

2. Increase user satisfaction

3. Increase user loyalty

4. Sell different items

5. Understand user needs

The above motivations have increased the number of e-commerce providers adopt recommender systems to their core of business. The users also need Recommender Systems if it can help their needs. It is important to find a balance between the two players and offer service of value to both.

There are number of ways you can use the recommender systems, the most obvious is to suggest useful items to the user. The other ways are more opportunistic in the sense [8], for example, a search engine which is also a kind of recommender system, is used to locate documents are relevant to the user needs/query. This system can also be used to check the importance of the web page, discover word combinations and word usages in a collection of web pages, etc. Recommending items to users as a ranked list along with some predictions on how much each item is relevant to the users preferences; Given a context, emphasize on the users long-term interests, for example, a TV recommender system might annotate users preferred TV shows in the TV guide; Instead of merely recommending one best item that fit the needs focusing on recommending a sequence of items could improve user experience, for example, suggesting a sequence music choices from the same genre depending on the user query; Recommending a group of items that work well together rather individually, for example, planning a travel would involve carefully selecting the destination in accordance to the time and various accommodations along the way; The user might browse items without an intention to buy, the catalog needs to rendered such that the users interests are matched against the items and recommendations must fall within the scope of the user's interests [9]; Sometimes the recommender systems are not that reliable in making good recommendations, some systems out there offer functions to test the system behavior in addition to obtain recommendations; The system must allow the users to update their interests, likes and dislikes. This is strictly necessary in providing personalized recommendations, if the system does not have access to the user's specific interests then provide only the recommendations similar to an "average" user;Some users may not care about recommendations and may want to contribute to the system. It is important the system provide facilities to users who care about expressing their interests through ratings, reviews, etc. This helps in maintaining user loyalty and the information goes a long way in contributing to the efficiency of the system; The recommender systems can also be fooled by malicious users who are hell bent on influencing certain products, some products are rated low for malicious reasons. The system should account for such ratings.

8.1.2 Data Sources for Recommender Systems

Recommender Systems are largely information processing systems, they gather various kinds of information about the items in order to build recommendations. Data is usually about the *items* and *users*. Whether this data can be exploited depends on the recommendation technique used. In general there are recommendation techniques that are knowledge independent, for example, basic recommendations based on user ratings which is simply a data query listing. While some recommendation techniques are knowledge dependent, for example, using social interactions of users.

The data used by recommender systems are usually of three kinds: *items*, *users*, and *transactions*.

Items are objects that are recommended. Items are characterized based on value, utility, and complexity. The value of the item is positive or negative based on its usefulness to the user. There are two types of costs that affect recommending an item. *Cognitive Cost* of searching for the item and *Monetory Cost* of the item itself. For example, there is cognitive associated in searching for news even if the user is not paying for that news. If this piece of news is selected then the cost is dominated by the benefit of acquiring the news, but if the item is rejected then the net value of the item/news piece is negative for that user. In case of cars, financial investments, the actual Monetory Cost of the item is used in the recommendations. In complex items like movie recommendation systems, the genre(comedy, action, etc), the director, cast are used to learn how the movie depends on its features [10].

Users may have different goals and characteristics. A range of data points about the users are captured to personalize the recommendations. This information can be structured in various ways depending on the recommendation technique used. In collaborative filtering method the users are stored as a list containing user ratings on different items. In a demography dependent recommender system age, gender, profession and education are modeled [11]. Recommender systems can also be defined as a tool that generates suggestions based on the user models, personalized suggestions are not possible without user models [12]. Users can also be modeled by their behavior patterns, for example, site browsing history, travel search history, etc [13]. Morever trust levels of different users can also be captured so that recommendations can be made based on trusted users.

Transactions indicate recorded interaction between a user and a recommender system. They are log-like that store important information about the interaction, for instance, it may contain the selected item and context of the recommendation, (e.g., query). Transactions can also include feedbacks like item ratings. These can be captured in number of ways, few are given below [14]:

- Numerical ratings such as star system, eg: Amazon 5 stars meaning the highest rating.

- Opinion ratings such as strongly agree, agree, neutral, disagree, strongly disagree

- Binary ratings that model the items if the user likes or dislikes it.

- Unary ratings indicate if the user has observed, purchased, or rated an item. The absence of which indicates there is no relationship between the user and item.

While collecting implicit ratings, the systems aims to observe the opinions based on user interactions. For example, if a user queries for "Distributed Systems", Amazon gives a list of items that match the query. If one of the item is selected to view further information, then we can assume that the user is interested in such an item. In transaction-based systems, the recommender systems collect several request-responses and learns from this data to provide future responses that seem to be more refined.

8.1.3 Techniques used in Recommender Systems

The Recommender System must predict the worth of each item before identifying useful items. Predicting the worth means to determine the utility of each item or compare the utility of items, then use this comparison for recommendation. The prediction step may not explicitly be a part of the algorithm but this can still be used in recommending items generally. For example, in order to recommend songs to users when there is lack of info about the users preferences, a popular song, item with high utility can be selected instead of a random song. This will have greater acceptance to a generic user. Some systems, knowledge-based, make recommendations by applying certain heuristics to hypothesize an item's use. The items utility can be Boolean and therefore the system will determine whether the item is useful or not. If there is some knowledge about the user, items and other users who have previously received recommendations, the system will utilize this knowledge with an appropriate algorithm to generate various predictions [2].

The items utility for an user can also depend on other variables, generally termed as "contextual" [15], for example, the domain knowledge of an user, time of user request, location-based requests. The recommendations need to adapt to these additional details and naturally the recommendations become harder and harder to achieve correct estimations.

Content-based: Here the items are recommended based on the similarity measure of the features between them. For example, if a user positively rates a movie in a comedy genre, the system can learn to recommend other movies in this genre. A content-based recommendation engine analyzes documents and descriptions of items rated by users and builds a model around the features of items rated by a user. The recommendation process is usually a way to match user attributes to content attributes. If a model accurately reflects user preferences it is important to assess the effectiveness of information access. Overview of Content-based system consists of:

- **Analyzer:** A pre-processing step that extracts structured or relevant information. Data items are analyzed by feature extraction techniques and convert the information from original format to target format. For example, converting documents to keyword vectors.

- **Learner:** This component constructs a model by generalizing user preferences data. The generalization techniques are adopted from machine learning [16] which are used to develop a model of user interests by looking at items which are liked or disliked.

- **Filter:** This component matches the user model with items and suggests the relevant items. Similarity metrics are used for judgment, resulting in a ranked list of relevant items.

Advantages of Content-based systems are:

- *User Dependence:* It relies primarily on the ratings provided by active users to build a model. While in collaborative filtering, the "nearest neighbor" principle is used in recommendation where the user with similar tastes ratings are recommended.

- *Transparency:* The working of the recommender system can be determined by listing the features used in creating the model. Those features are indicative of the trustworthy factor of the system.

- *New Item:* They can recommend items without any ratings, while the collaborative filtering method suffers from first-rater problem.

Disadvantages are:

- *Limited Analysis:* The number of features that can be used in recommendation is limited and domain knowledge is important for efficient recommendations. It is difficult for content-based systems to provide suggestions if there are less features to discriminate against between items. It cannot be used alone to provide a good recommendation as there are features apart from item feature that affect the recommendation.

- *Over-Specialization:* The items recommended are usually the ones that are rated highly against the user preferences. This results in lack of novelty in the recommendations, the suggestions are usually top rated items.

- *New User:* Before the system can provide good suggestions enough ratings has to be collected otherwise it is not possible to provide reliable recommendations.

Collaborative filtering: This approach recommends to the active user items that other users with similar interests have liked in the past. A simple way to determine similarity is by the similarity in the rating history of the users. It is sometimes called as "people-to-people" correlation which is one of the most popular recommender system techniques. The hypothesis for this method is that the rating of a new item i by user u will be similar to another user v, if both u and v have rated items in a similar way.

Collaborative filtering addresses a drawback of content-based systems where items whose details can be difficult to obtain can still be recommended if other users have provided feedback on it. It depends on the quality of items as evaluated by other users rather than the item's content. Collaborative filtering methods can be classified broadly in to two categories: *neighborhood* and *model* methods. In neighborhood based models, also called heuristic-based or memory-based, the predictions are made directly from the user-item ratings. For example, Grouplens [17], the interest of an user u for an item i is evaluated using the ratings of other users, *neighbors*, that have similar rating history. This is termed as user-based neighbor method. In item-based neighbor method, the rating of an item i by an user u is predicted from the rating pattern of the user u for similar items. Two items are similar if they have been rated in the same way with respect to a user [18].

In model-based approaches, the ratings are directly used in learning a predictive model. The general approach is to model the features of the users and items in to the system like the user preferences and item categories. There are numerous model-based approaches like Bayesian Clustering [19], Latent Semantic Analysis [20], Maximum Entropy [21], etc.

Demographic: The demographic profile of the user is critical to these kind of systems, providing different recommendations for different demographic classes. For example, users are directed to websites based on their language or country, recommendations are further customized based on the age of the user. Although these approaches have been commonly used with other techniques, very little research is done in demographic recommender systems [22].

Knowledge-based: The recommendation of items are based on the domain specific information about how items address the needs of users. These are case-based [23] - A similarity function estimates the degree of match between user needs and item recommendations. Another class of knowledge-based recommender system is similar to case-based with respect to the knowledge used, but differs in how the solutions are calculated. Explicit rules are provided to determine how the item features are related to the user preferences. They are good at the beginning of deployment but can easily be by-passed if you can overcome the rules.

Community-based: This type of recommender systems use the preferences of the users close circle / friends to suggest items. It is evident that people tend to trust the recommendation suggested by their friends rather than anonymous individuals [24]. The growing popularity of social networks like Twitter, Facebook has increased the usage of community-based systems. This system dwells on the social relationships of the active user and the preferences of users' friends. The rise in social networks has eased the data collection component in these systems. This area is still nascent, in the sense the result of using these systems have been mixed, where the results in few cases have been no better than traditional recommenders except in cases of highly rated items. Traditional recommenders along with social-network data can yield better results [25].

8.2 Implementation Details

Here, we implement an item-based recommender system using Scalding and Spark on a dataset of movies to suggest movies based on similar ratings.

Dataset We choose the MovieLens Dataset [28] The dataset contains two files

- **ua.base:** The data is randomly ordered. This is a tab separated list of *user id | item id | rating | timestamp.*The time stamps are unix seconds since 1/1/1970 UTC

- **u.item:** This file contains entries separted by "|". The entries are: *movie id | movie title | release date | video release date |* IMDb URL | unknown | *Action | Adventure | Animation | Children's | Comedy | Crime | Documentary | Drama | Fantasy | Film-Noir | Horror | Musical | Mystery | Romance | Sci-Fi | Thriller | War | Western.* The last 19 fields are the genres, a 1 indicates the movie is of that genre, a 0 indicates it is not.

ua.base, containing `userID`, `movieID` and `rating`, **u.item** containing `movieID` and `movieName`.

We have striped the Movielens dataset to include only the `movie id` and the `movie title`.

Table 8.1: u.item Dataset: Movie Id's and Movie Titles.

Movie ID	Movie Title
1	Toy Story (1995)
2	GoldenEye (1995)
3	Four Rooms (1995)
4	Get Shorty (1995)
5	Copycat (1995)
6	Shanghai Triad (1995)
7	Twelve Monkeys (1995)
8	Babe (1995)
9	Dead Man Walking (1995)
10	Richard III (1995)
11	Seven (Se7en) (1995)
12	Usual Suspects, The (1995)
13	Mighty Aphrodite (1995)
14	Postino, Il (1994)
15	Mr. Holland's Opus (1995)

Sample data from the Movielens dataset is shown in Table 8.1 8.2

Table 8.2: ua.base dataset

user Id	Movie Id	rating	Timestamp
1	1	5	874965758
1	2	3	876893171
1	3	4	878542960
1	4	3	876893119
1	5	3	889751712
1	6	5	887431973
1	7	4	875071561
1	8	1	875072484
1	9	5	878543541
1	10	3	875693118
1	11	2	875072262
1	12	5	878542960
1	13	5	875071805
1	14	5	874965706
1	15	5	875071608

Concept: Imagine you are an owner of an online movie business and you want to generate movie recommendations based on the user preferences. You introduce a rating system (1 to 5 stars). We use these ratings as the user preferences and the user to movie mapping to recommend future movies.

You want to be able to calculate how similar a pair of movies are, or that if a user watches a movie *Lion King* you can recommend him/her to watch *Toy Story*. How do we determine the similarity between two entities?

Solution: Correlation

- For every pair of the movies A and B find all the people who rated A and B

- Use these ratings to form a Movie A vector and Movie B vector.

- Calculate the Correlation between the two vectors.

- Whenever someone watches a particular movie that has been rated you can recommend a new movie which has close correlation to it.

8.2.1 Spark Implementation

Initialization of the Spark Project: Inorder for you to compile and execute Spark based programs you need to create a Spark project. The steps involved are:

- Create a directory structure of `src/main/scala`

- Create a `project-name.sbt` file also called the scala build too file, that is used to declare project dependencies, track project versions, etc.
- Create the Scala Spark Class in the `scala` folder path created earlier.
- Use `sbt package` command to compile and build the project.
- Use the following code sample to submit the program to the Spark Cluster Master.

```
spark-submit --class <Class-Name> --master <cluster Url>
<path/to/project.jar>
```

- replace `<Class-Name>` with the Spark Class name you have used in the program

- replace `cluster Url` with the path and port of the Spark Master, typically it is `spark://hostname:7077`.

- `path/to/project.jar` is replaced typically by the jar target/scala-2.10/ project.jar.

Step 1: Create a Spark Configuration and a Spark Context which acts as a entry point to the Spark engine. Spark context provides the base API's required for performing Spark related actions.

```
import org.apache.spark.SparkContext
import org.apache.spark.SparkContext._
import org.apache.spark.SparkConf

object Recommend {
    def main(args: Array[String]) {
      val conf = new SparkConf()
            .setAppName("RecommendMovies")
      val sc = new SparkContext(conf)
    }
}
```

Step 2: Read the dataset `u.item` which contains the tab seperated information about the Movie and its Genre, refer Table 8.1.

The input is split on the \t to obtain tuples (MovieId, MovieTitle).

```
import org.apache.spark.SparkContext
import org.apache.spark.SparkContext._
import org.apache.spark.SparkConf

object Recommend {
    def main(args: Array[String]) {
      val conf = new SparkConf()
            .setAppName("RecommendMovies")

      val sc = new SparkContext(conf)
```

```
val movies = sc.textFile("u.item")
    .map(line => {
        val fields = line.split("\t")
        (fields(0).toInt, fields(1))
    })
}
}
```

Step 3: The Array of Tuples obtained in the Step 2 is collected as map for future mapping based on the *MovieId* to return the *MovieTitle*. This will be used for pretty printing the output after the similarity calculation.

We use the Spark API `collectAsMap()`. The API definition looks like this:

```
def collectAsMap(): Map[K, V]
```

`collectAsMap` returns a HashMap[K,V] where K is the `MovieId` and V is the `MovieTitle`.

```
import org.apache.spark.SparkContext
import org.apache.spark.SparkContext._
import org.apache.spark.SparkConf

object Recommend {
    def main(args: Array[String]) {
        val conf = new SparkConf()
            .setAppName("RecommendMovies")

        val sc = new SparkContext(conf)

        val movies = sc.textFile("u.item")
            .map(line => {
                val fields = line.split("\t")
                (fields(0).toInt, fields(1))
            })

        val moviesName = movies.collectAsMap()

    }
}
```

Step 4: Read the `ua.base` dataset and extract the `userID`, `MovieID` and the `rating` fields.

The input is read as Strings, hence need to be converted to Integer values for arithmetic expressions.

```scala
import org.apache.spark.SparkContext
import org.apache.spark.SparkContext._
import org.apache.spark.SparkConf

object Recommend {
  def main(args: Array[String]) {
    val conf = new SparkConf()
        .setAppName("RecommendMovies")

    val sc = new SparkContext(conf)

    val movies = sc.textFile("u.item")
      .map(line => {
        val fields = line.split("\t")
        (fields(0).toInt, fields(1))
      })

    val moviesName = movies.collectAsMap()

    val ratings = sc.textFile("ua.base")
      .map(line => {
        val fields = line.split("\t")
        (fields(0).toInt,
        fields(1).toInt,
        fields(2).toInt)
      })
  }
}
```

Step 5: Calculate the number of raters for each movie. The `movieID` is the key and the `numRaters` is the value.

We use the `groupBy` Spark API, which has the following definition, It returns an RDD of grouped items where each group consits of a key and sequence of values mapped to that key. In our case the Key is the `movieID` and the values inherently become the `rating`.

```scala
def groupBy[K](f: T => K): RDD[(K, Iterable[T])]
```

Once we have the following arrangement as a result from the previous statement, $(movieID, [rating_1, rating_2, ..., rating_n])$. we need to find the size of the Value vector mapped to a `movieID`.

We use the Spark `map` API to iterate over each movie group and find the size of its associated ratings vector.

```
ratings.map(grouped => (grouped._1, grouped._2.size))
```

The code looks like below:

```scala
import org.apache.spark.SparkContext
import org.apache.spark.SparkContext._
import org.apache.spark.SparkConf

object Recommend {
  def main(args: Array[String]) {
    val conf = new SparkConf()
      .setAppName("RecommendMovies")

    val sc = new SparkContext(conf)

    val movies = sc.textFile("u.item")
        .map(line => {
          val fields = line.split("\t")
          (fields(0).toInt, fields(1))
        })

    val moviesName = movies.collectAsMap()

    val ratings = sc.textFile("ua.base")
      .map(line => {
        val fields = line.split("\t")
        (fields(0).toInt,
        fields(1).toInt,
        fields(2).toInt)
      })

    val numRatersPerMovie = ratings
      .groupBy(tup => tup._2)
      .map(grouped =>
            (grouped._1,
            grouped._2.size))
      .saveAsTextFile("numRatersPerMovie.txt")
  }
}
```

Sample Output: contains the `movieId` and the `numRaters`.

Table 8.3: Tuple of *movieId* and *numRaters*

Tuple(movieId, numRaters)
(1084,17)
(454,13)
(1410,4)
(772,46)
(752,36)
(586,32)
(428,115)
(1328,6)
(464,26)
(14,161)
(466,49)
(1040,22)
(912,7)
(1338,4)
(1494,1)
(1336,9)

Step 6: Now we have calculated the number of raters for each movie, we need to add this to the original list of movie ratings. We perform a `join` operation over the `movieID`.

Spark `join` API definition looks like this; It returns an RDD containing all pairs of elements with matching keys in *this* RDD and *other* RDD. Each pair of elements will be returned as a $(k, (v_1, v_2))$ tuple, where (k, v_1) is in *this* and (k, v_2) is in *other*. Uses the given Partitioner to partition the output RDD.

```
def join[W](other: RDD[(K, W)],
        partitioner: Partitioner): RDD[(K, (V, W))]
```

Once we obtained the joined results we need to flatten the result so that we have clear, distinct fields for futher calculations. Spark provides a `flatMap` API that does exactly this:

It returns an RDD by first applying a function on all its elements and then flattening the result.

```
def flatMap[U: ClassTag]
            (f: T => TraversableOnce[U]): RDD[U]
```

The function inside `flatMap` is a simple projection of the chosen fields. The fields chosen are `userID`, `movieID`, `rating`, `numOfRatings`

```
import org.apache.spark.SparkContext
import org.apache.spark.SparkContext._
import org.apache.spark.SparkConf

object Recommend {
  def main(args: Array[String]) {
```

```
val conf = new SparkConf()
  .setAppName("RecommendMovies")

val sc = new SparkContext(conf)

val movies = sc.textFile("u.item")
  .map(line => {
    val fields = line.split("\t")
    (fields(0).toInt, fields(1))
  })

val moviesName = movies.collectAsMap()

val ratings = sc.textFile("ua.base")
  .map(line => {
    val fields = line.split("\t")
    (fields(0).toInt,
     fields(1).toInt,
     fields(2).toInt)
  })

val numRatersPerMovie = ratings
  .groupBy(tup => tup._2)
  .map(grouped =>
        (grouped._1,
         grouped._2.size))

val ratingsWithSize = ratings
  .groupBy(tup => tup._2)
  .join(numRatersPerMovie)
  .flatMap(joined => {
    joined._2._1.map(f => (
                    f._1,
                    f._2,
                    f._3,
                    joined._2._2))
  })
  .saveAsTextFile("ratingsWithSize.txt")
  }
}
```

Step 7: In order to calculate the correlation of two movies we need to create a combination such that $(userID, (movie_1, movie_2))$. The combination should not contain same movies as pairs, as in $(userID, (movie_1, movie_1))$ which is possible during a self-join operation. To avoid this, we filter the combinations such that $movie_1 < movie_2$, ensuring we get unique pairs.

First, we create a duplicate RDD of the ratings for performing a self-join operation. Secondly, perform the join operation over the `userID`. Thirdly, filter the movie combinations such that $movieID_1 < movieID_2$.

```scala
import org.apache.spark.SparkContext
import org.apache.spark.SparkContext._
import org.apache.spark.SparkConf

object Recommend {
  def main(args: Array[String]) {
  val conf = new SparkConf()
    .setAppName("RecommendMovies")

  val sc = new SparkContext(conf)

  val movies = sc.textFile("u.item")
    .map(line => {
      val fields = line.split("\t")
      (fields(0).toInt, fields(1))
    })

  val moviesName = movies.collectAsMap()

  val ratings = sc.textFile("ua.base")
   .map(line => {
     val fields = line.split("\t")
     (fields(0).toInt,
     fields(1).toInt,
     fields(2).toInt)
   })

   val numRatersPerMovie = ratings
     .groupBy(tup => tup._2)
     .map(grouped =>
            (grouped._1,
            grouped._2.size))

   val ratingsWithSize = ratings
     .groupBy(tup => tup._2)
     .join(numRatersPerMovie)
     .flatMap(joined => {
       joined._2._1.map(f => (
                     f._1,
                     f._2,
                     f._3,
                     joined._2._2))
     })

   val ratings2 = ratingsWithSize.keyBy(tup => tup._1)

   val ratingPairs =
```

```
            ratingsWithSize
            .keyBy(tup => tup._1)
            .join(ratings2)
            .filter(f => f._2._1._2 < f._2._2._2)
            .saveAsTextFile("ratingPairs")

    }
}
```

Sample Output: Movie Pairs for calculating the correlation.

Table 8.4: Movie Pairs

(userID, ((ratings for $movie_i$),(ratings for $movie_j$)))
(68,((68,282,1,206),(68,288,4,386)))
(68,((68,282,1,206),(68,286,5,400)))
(68,((68,282,1,206),(68,926,1,94)))
(68,((68,282,1,206),(68,1028,4,133)))
(68,((68,282,1,206),(68,458,1,83)))
(68,((68,282,1,206),(68,596,2,115)))
(68,((68,282,1,206),(68,471,3,195)))
(68,((68,282,1,206),(68,475,5,222)))
(68,((68,282,1,206),(68,763,1,138)))

Step 8: We identify movies as a vector of ratings. To calculate the correlation we need:

- Dot product of ratings : Multiply the Ratings of $movie_i$ and Ratings of $movie_j$.

- Rating norms

- Squared ratings of $movie_i$

- Squared ratings of $movie_j$

- Sum of ratings of $movie_i$

- Sum of ratings of $movie_j$

- size of each movie vector

To calculate the rating dot product, rating squares we use the Spark map API that
iterates over each item and performs a function on it.

```scala
val vectorCalcs = ratingPairs
.map(data => {
  val key = (data._2._1._2, data._2._2._2)
  val stats =
    (data._2._1._3 * data._2._2._3,  // rating 1 * rating 2
     data._2._1._3,                   // rating movie 1
     data._2._2._3,                   // rating movie 2
     math.pow(data._2._1._3, 2),      // square of rating movie 1
     math.pow(data._2._2._3, 2),      // square of rating movie 2
     data._2._1._4,                   // number of raters movie 1
     data._2._2._4)                   // number of raters movie 2
  (key, stats)
})
```

Here the map function operates on the data and outputs:

$$key : (movie_i, movie_j),$$

$$stats : (ratings_i * ratings_j, rating_i, rating_j, rating_i^2, rating_j^2, numRaters_i, numRaters_j)$$

```scala
import org.apache.spark.SparkContext
import org.apache.spark.SparkContext._
import org.apache.spark.SparkConf

object Recommend {
  def main(args: Array[String]) {
  val conf = new SparkConf()
    .setAppName("RecommendMovies")

  val sc = new SparkContext(conf)

  val movies = sc.textFile("u.item")
    .map(line => {
      val fields = line.split("\t")
      (fields(0).toInt, fields(1))
    })

  val moviesName = movies.collectAsMap()

  val ratings = sc.textFile("ua.base")
    .map(line => {
      val fields = line.split("\t")
      (fields(0).toInt,
       fields(1).toInt,
       fields(2).toInt)
    })

    val numRatersPerMovie = ratings
      .groupBy(tup => tup._2)
      .map(grouped =>
```

```
                    (grouped._1,
                    grouped._2.size))

    val ratingsWithSize = ratings
      .groupBy(tup => tup._2)
      .join(numRatersPerMovie)
      .flatMap(joined => {
        joined._2._1.map(f => (
                          f._1,
                          f._2,
                          f._3,
                          joined._2._2))
      })

    val ratings2 = ratingsWithSize.keyBy(tup => tup._1)

    val ratingPairs =
        ratingsWithSize
        .keyBy(tup => tup._1)
        .join(ratings2)
        .filter(f => f._2._1._2 < f._2._2._2)

val vectorCalcs = ratingPairs
    .map(data => {
      val key = (data._2._1._2, data._2._2._2)
      val stats =
        (data._2._1._3 * data._2._2._3,
          data._2._1._3,
          data._2._2._3,
          math.pow(data._2._1._3, 2),
          math.pow(data._2._2._3, 2),
          data._2._1._4,
          data._2._2._4)
      (key, stats)
    })
  }
}
```

Step 9: Correlation of (X, Y) can be expressed as [32]:

$$Corr(X,Y) = \frac{n\sum xy - \sum x \sum y}{\sqrt{n\sum x^2 - (\sum x)^2}\sqrt{n\sum y^2 - (\sum y)^2}}$$

According to which we need:

- sum(X · Y)
- sum(X)
- sum(Y)

- $sum(X^2)$

- $sum(Y^2)$

This can be implemeted using Spark's `groupByKey()` API that has the definition as below, It returns an RDD which is grouped by key into a single sequence.

```
def groupByKey(): RDD[(K, Iterable[V])]
```

Applying a Spark map function to the Iterable from the result to calculate the sum of various Fields.

```
ratingsPairs.groupByKey()
 .map(data => {
      val key = data._1
      val vals = data._2
      val size = vals.size
      val dotProduct = vals.map(f => f._1).sum
      val ratingSum = vals.map(f => f._2).sum
      val rating2Sum = vals.map(f => f._3).sum
      val ratingSq = vals.map(f => f._4).sum
      val rating2Sq = vals.map(f => f._5).sum
      val numRaters = vals.map(f => f._6).max
      val numRaters2 = vals.map(f => f._7).max
     (key, (size, dotProduct, ratingSum, rating2Sum,
       ratingSq, rating2Sq, numRaters, numRaters2))
   })
```

Note: `numRaters` and `numRaters2` are subjected to a `max` function, but there is no effect as they are a sequence of the same number in either case, max([a,a,a]) is a. Such a function is applied just to ensure the fields are visible for similarity calculations.

```
import org.apache.spark.SparkContext
import org.apache.spark.SparkContext._
import org.apache.spark.SparkConf

object Recommend {
 def main(args: Array[String]) {
 val conf = new SparkConf()
   .setAppName("RecommendMovies")

 val sc = new SparkContext(conf)

 val movies = sc.textFile("u.item")
   .map(line => {
     val fields = line.split("\t")
     (fields(0).toInt, fields(1))
   })

 val moviesName = movies.collectAsMap()

 val ratings = sc.textFile("ua.base")
   .map(line => {
```

```
     val fields = line.split("\t")
     (fields(0).toInt,
     fields(1).toInt,
     fields(2).toInt)
  })

  val numRatersPerMovie = ratings
     .groupBy(tup => tup._2)
     .map(grouped =>
            (grouped._1,
            grouped._2.size))

  val ratingsWithSize = ratings
     .groupBy(tup => tup._2)
     .join(numRatersPerMovie)
     .flatMap(joined => {
       joined._2._1.map(f => (
                       f._1,
                       f._2,
                       f._3,
                       joined._2._2))
     })

  val ratings2 = ratingsWithSize.keyBy(tup => tup._1)

  val ratingPairs =
       ratingsWithSize
       .keyBy(tup => tup._1)
       .join(ratings2)
       .filter(f => f._2._1._2 < f._2._2._2)

val vectorCalcs = ratingPairs
   .map(data => {
     val key = (data._2._1._2, data._2._2._2)
     val stats =
       (data._2._1._3 * data._2._2._3,
         data._2._1._3,
         data._2._2._3,
         math.pow(data._2._1._3, 2),
         math.pow(data._2._2._3, 2),
         data._2._1._4,
         data._2._2._4)
     (key, stats)
   })
   .groupByKey()
   .map(data => {
     val key = data._1
     val vals = data._2
     val size = vals.size
     val dotProduct = vals.map(f => f._1).sum
     val ratingSum = vals.map(f => f._2).sum
     val rating2Sum = vals.map(f => f._3).sum
     val ratingSq = vals.map(f => f._4).sum
     val rating2Sq = vals.map(f => f._5).sum
```

```
      val numRaters = vals.map(f => f._6).max
      val numRaters2 = vals.map(f => f._7).max
      (key, (size, dotProduct, ratingSum, rating2Sum,
        ratingSq, rating2Sq, numRaters, numRaters2))
    })
  }
}
```

Step 10: We have now calculated all the necessary variables for the calculation of Correlation(X,Y). We simple apply the equation on these values.

We define a function `correlation`

```
def correlation(size : Double,
                dotProduct : Double,
                ratingSum : Double,
                rating2Sum : Double,
                ratingNormSq : Double,
                rating2NormSq : Double) = {

val numerator = size*dotProduct - ratingSum*rating2Sum
val denominator =
scala.math.sqrt(size*ratingNormSq - ratingSum*ratingSum) *
scala.math.sqrt(size*rating2NormSq - rating2Sum*rating2Sum)

numerator / denominator

}
```

The resulting RDD is mapped to the above correlation function.

```
import org.apache.spark.SparkContext
import org.apache.spark.SparkContext._
import org.apache.spark.SparkConf

object Recommend {
 def main(args: Array[String]) {
 val conf = new SparkConf()
    .setAppName("RecommendMovies")

 val sc = new SparkContext(conf)

 val movies = sc.textFile("u.item")
   .map(line => {
     val fields = line.split("\t")
     (fields(0).toInt, fields(1))
   })

 val moviesName = movies.collectAsMap()

 val ratings = sc.textFile("ua.base")
  .map(line => {
    val fields = line.split("\t")
    (fields(0).toInt,
```

```
      fields(1).toInt,
      fields(2).toInt)
  })

  val numRatersPerMovie = ratings
    .groupBy(tup => tup._2)
    .map(grouped =>
            (grouped._1,
             grouped._2.size))

  val ratingsWithSize = ratings
    .groupBy(tup => tup._2)
    .join(numRatersPerMovie)
    .flatMap(joined => {
      joined._2._1.map(f => (
                        f._1,
                        f._2,
                        f._3,
                        joined._2._2))
    })

  val ratings2 = ratingsWithSize.keyBy(tup => tup._1)

  val ratingPairs =
      ratingsWithSize
      .keyBy(tup => tup._1)
      .join(ratings2)
      .filter(f => f._2._1._2 < f._2._2._2)

val vectorCalcs = ratingPairs
  .map(data => {
    val key = (data._2._1._2, data._2._2._2)
    val stats =
      (data._2._1._3 * data._2._2._3,
        data._2._1._3,
        data._2._2._3,
        math.pow(data._2._1._3, 2),
        math.pow(data._2._2._3, 2),
        data._2._1._4,
        data._2._2._4)
    (key, stats)
  })
  .groupByKey()
  .map(data => {
    val key = data._1
    val vals = data._2
    val size = vals.size
    val dotProduct = vals.map(f => f._1).sum
    val ratingSum = vals.map(f => f._2).sum
    val rating2Sum = vals.map(f => f._3).sum
    val ratingSq = vals.map(f => f._4).sum
    val rating2Sq = vals.map(f => f._5).sum
    val numRaters = vals.map(f => f._6).max
    val numRaters2 = vals.map(f => f._7).max
```

```
      (key, (size, dotProduct, ratingSum, rating2Sum,
        ratingSq, rating2Sq, numRaters, numRaters2))
    })

 val similarities = vectorCalcs
    .map(fields => {
      val key = fields._1
      val (size, dotProduct, ratingSum, rating2Sum,
        ratingNormSq, rating2NormSq, numRaters,
        numRaters2) = fields._2
      val corr = correlation(size, dotProduct, ratingSum,
              rating2Sum, ratingNormSq, rating2NormSq)
      (key, corr)
    })
    .saveAsTextFile("withcorr.txt")
 }

def correlation(size : Double,
                dotProduct : Double,
                ratingSum : Double,
                rating2Sum : Double,
                ratingNormSq : Double,
                rating2NormSq : Double) = {

val numerator = size*dotProduct - ratingSum*rating2Sum
val denominator =
scala.math.sqrt(size*ratingNormSq - ratingSum*ratingSum) *
scala.math.sqrt(size*rating2NormSq - rating2Sum*rating2Sum)

numerator / denominator
 }
 }
```

Step 11: Apart from correlation function there are other similarity measures such as:

- **Cosine Similarity** : is a measure of the similarity between two vectors on their inner products that measures cosine of the anlge between them [30]. The cosine of two vectors can be derived from euclidean dot product.

General equation for two vectors:

$$x \cdot y = \|x\| \, \|y\| \, cos\theta$$

Given two vectors X and Y, the similarity is measured by:

$$cos(\theta) = \frac{X \cdot Y}{\|X\| \|Y\|}$$

$$= \frac{\sum_{i=1}^{n} X_i \times Y_i}{\sqrt{\sum_{i=1}^{n}(X_i)^2} \times \sqrt{\sum_{i=1}^{n}(Y_i)^2}}$$

```
def cosineSimilarity(dotProduct : Double,
                     ratingNorm : Double,
                     rating2Norm : Double) = {

  dotProduct / (ratingNorm * rating2Norm)
}
```

- **Regularized Correlation**: This is applied to reduce noise in scenarios where there might be few raters in common. This is done by:

```
val virtualCount = 10
val priorCorrelation = 0

def regularizedCorrelation(size : Double,
                           dotProduct : Double,
                           ratingSum : Double,
                           rating2Sum : Double,
                           ratingNormSq : Double,
                           rating2NormSq : Double,
                           virtualCount : Double,
                           priorCorrelation : Double) = {

  val unregularizedCorrelation = correlation(size,
                                             dotProduct,
                                             ratingSum,
                                             rating2Sum,
                                             ratingNormSq,
                                             rating2NormSq)

  val w = size / (size + virtualCount)

  w * unregularizedCorrelation + (1 - w) * priorCorrelation
}
```

- **Jaccard Similarity:** It is a method to determine the similarity or diversity of two sample sets. [31] It is expressed as:

$$J(X,Y) = \frac{|X \cap Y|}{|X \cup Y|}$$

```
def jaccardSimilarity(commonRaters : Double,
                      raters1 : Double,
                      raters2 : Double) = {
  val union = raters1 + raters2 - commonRaters
```

```
    commonRaters / union
  }
```

Final version of the code looks like below:

```scala
import org.apache.spark.SparkContext
import org.apache.spark.SparkContext._
import org.apache.spark.SparkConf

object Recommend {
  def main(args: Array[String]) {
  val conf = new SparkConf()
     .setAppName("RecommendMovies")

  val sc = new SparkContext(conf)

  val movies = sc.textFile("u.item")
    .map(line => {
      val fields = line.split("\t")
      (fields(0).toInt, fields(1))
    })

  val moviesName = movies.collectAsMap()

  val ratings = sc.textFile("ua.base")
   .map(line => {
     val fields = line.split("\t")
     (fields(0).toInt,
     fields(1).toInt,
     fields(2).toInt)
   })

  val numRatersPerMovie = ratings
    .groupBy(tup => tup._2)
    .map(grouped =>
          (grouped._1,
          grouped._2.size))

  val ratingsWithSize = ratings
    .groupBy(tup => tup._2)
    .join(numRatersPerMovie)
    .flatMap(joined => {
      joined._2._1.map(f => (
                    f._1,
                    f._2,
                    f._3,
                    joined._2._2))
    })

  val ratings2 = ratingsWithSize.keyBy(tup => tup._1)

  val ratingPairs =
     ratingsWithSize
     .keyBy(tup => tup._1)
```

```scala
      .join(ratings2)
      .filter(f => f._2._1._2 < f._2._2._2)

val vectorCalcs = ratingPairs
    .map(data => {
      val key = (data._2._1._2, data._2._2._2)
      val stats =
        (data._2._1._3 * data._2._2._3,
          data._2._1._3,
          data._2._2._3,
          math.pow(data._2._1._3, 2),
          math.pow(data._2._2._3, 2),
          data._2._1._4,
          data._2._2._4)
      (key, stats)
    })
    .groupByKey()
    .map(data => {
      val key = data._1
      val vals = data._2
      val size = vals.size
      val dotProduct = vals.map(f => f._1).sum
      val ratingSum = vals.map(f => f._2).sum
      val rating2Sum = vals.map(f => f._3).sum
      val ratingSq = vals.map(f => f._4).sum
      val rating2Sq = vals.map(f => f._5).sum
      val numRaters = vals.map(f => f._6).max
      val numRaters2 = vals.map(f => f._7).max
      (key, (size, dotProduct, ratingSum, rating2Sum,
       ratingSq, rating2Sq, numRaters, numRaters2))
    })

  val similarities = vectorCalcs
      .map(fields => {
        val key = fields._1
        val (size, dotProduct, ratingSum, rating2Sum,
             ratingNormSq, rating2NormSq, numRaters,
             numRaters2) = fields._2
        val corr = correlation(size, dotProduct, ratingSum,
                        rating2Sum, ratingNormSq, rating2NormSq)
        val regCorr = regularizedCorrelation(size, dotProduct,
                          ratingSum, rating2Sum, ratingNormSq,
                          rating2NormSq, PRIOR_COUNT,
                          PRIOR_CORRELATION)
        val cosSim = cosineSimilarity(dotProduct,
                          scala.math.sqrt(ratingNormSq),
                          scala.math.sqrt(rating2NormSq))
        val jaccard = jaccardSimilarity(size, numRaters,
                                        numRaters2)

        (key, (corr, regCorr, cosSim, jaccard))
      })
    .saveAsTextFile("withcorr.txt")
  }
```

```scala
def correlation(size : Double,
                dotProduct : Double,
                ratingSum : Double,
                rating2Sum : Double,
                ratingNormSq : Double,
                rating2NormSq : Double) = {

val numerator = size*dotProduct - ratingSum*rating2Sum
val denominator =
scala.math.sqrt(size*ratingNormSq - ratingSum*ratingSum) *
scala.math.sqrt(size*rating2NormSq - rating2Sum*rating2Sum)

numerator / denominator
}

def regularizedCorrelation(size : Double,
                           dotProduct : Double,
                           ratingSum : Double,
                           rating2Sum : Double,
                           ratingNormSq : Double,
                           rating2NormSq : Double,
                           virtualCount : Double,
                           priorCorrelation : Double) = {

  val unregularizedCorrelation = correlation(size,
                    dotProduct, ratingSum,
                    rating2Sum, ratingNormSq,
                    rating2NormSq)
  val w = size / (size + virtualCount)

  w*unregularizedCorrelation + (1-w)*priorCorrelation
}

def cosineSimilarity(dotProduct : Double,
                     ratingNorm : Double,
                     rating2Norm : Double) = {
  dotProduct / (ratingNorm * rating2Norm)
}

def jaccardSimilarity(commonRaters : Double,
                      raters1 : Double,
                      raters2 : Double) = {
  val union = raters1 + raters2 - commonRaters
  commonRaters / union
}
}
```

Output: Sample output for movie Star Wars (1977):

Table 8.5: Recommendation

Movie 1	Movie 2	Correlation	RCorrelation	Cosine	Jaccard
Star Wars (1977)	Empire Strikes Back, The (1980)	0.7419	0.7168	0.9888	0.5306
Star Wars (1977)	Return of the Jedi (1983)	0.6714	0.6539	0.9851	0.6708
Star Wars (1977)	Raiders of the Lost Ark (1981)	0.5074	0.4917	0.9816	0.5607
Star Wars (1977)	Meet John Doe (1941)	0.6396	0.4397	0.9840	0.0442
Star Wars (1977)	Love in the After-noon (1957)	0.9234	0.4374	0.9912	0.0181
Star Wars (1977)	Man of the Year (1995)	1.0000	0.4118	0.9995	0.0141

8.2.2 Scalding Implementation:

Step 1: Initialize a RecommendMovies class that extends the Scalding Job class, which implements the necessary I/O and Flow definitions that are required for the execution of the Scalding Job.

```
import com.twitter.scalding._
class RecommendMovies(args : Args) extends Job(args) {

// Code Goes Here.

}
```

Step 2: Read the input from as show in Table 8.2 which contains the movie details, userID, movieID, rating and timestamp.

```
import com.twitter.scalding._
class RecommendMovies(args : Args) extends Job(args) {
val INPUT_FILENAME = "ua.base"
val ratings = Tsv(INPUT_FILENAME)
  .read
  .mapTo((0, 1, 2) -> ('user, 'movie, 'rating)) {
    fields : (Int, Int, Double) => fields
  }
  .write(Tsv("ratings.tsv"))
}
```

Step 3: Find the number of Raters for each movie: We use Scalding Grouping functions to group over the movieID and find the size of the group.

```
import com.twitter.scalding._
class RecommendMovies(args : Args) extends Job(args) {
val ratings = Tsv("ua.base")
```

```
  .read
  .mapTo((0, 1, 2) -> ('user, 'movie, 'rating)) {
    fields : (Int, Int, Double) => fields
  }

val ratingsWithSize = ratings
  .groupBy('movie) { _.size('numRaters) }
  .write(Tsv("ratingsWithSize.tsv"))
}
```

Step 4: We need to join this grouped result against each movie for future similarity calculations. We use Scalding's `joinWithLarger` function to join two Pipes over `movieID`.

The API for `joinWithLarger` in Scalding is as follows, It takes arguments on the Joined Fields(fs), and *that* Cascading Pipe [29], along with the type of Join operation as `joiner`, defaults to InnerJoin, with the number of reducers.

```
def joinWithLarger(fs: (Fields, Fields),
                   that: Pipe,
                   joiner: Joiner = new InnerJoin,
                   reducers: Int = -1)
```

Note: It is important not to have any conflicting Field names when performing a Join. Hence, in our case we rename on the Joining Pipe's `movieID` to `movieX` and discard the duplicate.

```
import com.twitter.scalding._
class RecommendMovies(args : Args) extends Job(args) {
val ratings = Tsv("ua.base")
  .read
  .mapTo((0, 1, 2) -> ('user, 'movie, 'rating)) {
    fields : (Int, Int, Double) => fields
  }

val ratingsWithSize = ratings
  .groupBy('movie) { _.size('numRaters) }
  .write(Tsv("ratingsWithSize.tsv"))

val ratingsJoinWithSize = ratings
  .rename('movie -> 'movieX)
  .joinWithLarger('movieX -> 'movie, ratings)
  .discard('movieX)
  .write(Tsv("ratingsWithSize.tsv"))
}
```

Sample Output: The result has Fields `userID`, `movieID`, `ratings` and `numRatings` as shown in Table 8.6.

Step 5: Inorder to calculate the correlation between two movie vectors we need them as Fields for each occurance of movie rating.

Table 8.6: Movie ratings with number of Ratings

userID	movieID	ratings	numRatings
1	1	5.0	392
2	1	4.0	392
6	1	4.0	392
10	1	4.0	392
13	1	3.0	392
15	1	1.0	392
16	1	5.0	392
18	1	5.0	392
20	1	3.0	392
21	1	5.0	392
23	1	5.0	392
25	1	5.0	392

First we create a duplicate of *ratingJoinWithSize*.

```
import com.twitter.scalding._
class RecommendMovies(args : Args) extends Job(args) {
val ratings = Tsv("ua.base")
  .read
  .mapTo((0, 1, 2) -> ('user, 'movie, 'rating)) {
    fields : (Int, Int, Double) => fields
  }

val ratingsWithSize = ratings
  .groupBy('movie) { _.size('numRaters) }
  .write(Tsv("ratingsWithSize.tsv"))

val ratingsJoinWithSize = ratings
  .rename('movie -> 'movieX)
  .joinWithLarger('movieX -> 'movie, ratings)
  .discard('movieX)


val ratings2 = ratingsJoinWithSize
      .rename(('user, 'movie, 'rating, 'numRaters) ->
      ('user2, 'movie2, 'rating2, 'numRaters2))
      .write(Tsv("ratings2.tsv"))
}
```

Second, we create movie pairs to easily calculate the correlation between each pair. To achieve this we need to Join the ratings with itself in such a way that there are no duplicate pairs. The de-duplication can be done by checking if the $movieID_i <$ $movieID_j$ ensures unique pairs of movies.

We implement this by performing a JoinWithSmaller operation on the duplicate ratings2. The result of this must be filtered to remove the duplicates with $(movie_i, movie_i)$ combinations. Scalding provides a convenient function for filtering the data. The API looks like follows:

```
def filter[A](f: Fields)(fn: (A) => Boolean): Pipe
```

The `filter` function return a Pipe such that each entry of the Pipe is checked to satisfy the predicate defined by a function. In out case the function is ($movie_i <$ $movie_j$).

```
import com.twitter.scalding._
class RecommendMovies(args : Args) extends Job(args) {
val ratings = Tsv("ua.base")
  .read
  .mapTo((0, 1, 2) -> ('user, 'movie, 'rating)) {
    fields : (Int, Int, Double) => fields
  }

val ratingsWithSize = ratings
  .groupBy('movie) { _.size('numRaters) }
  .write(Tsv("ratingsWithSize.tsv"))

val ratingsJoinWithSize = ratings
  .rename('movie -> 'movieX)
  .joinWithLarger('movieX -> 'movie, ratings)
  .discard('movieX)


val ratings2 = ratingsJoinWithSize
      .rename(('user, 'movie, 'rating, 'numRaters) ->
      ('user2, 'movie2, 'rating2, 'numRaters2))

val ratingPairs =
    ratingsJoinWithSize
      .joinWithSmaller('user -> 'user2, ratings2)
      .filter('movie, 'movie2) {
        movies : (String, String) => movies._1 < movies._2}
      .project('movie, 'rating, 'numRaters, 'movie2,
            'rating2, 'numRaters2)
      .write(Tsv("ratingPairs.tsv"))
}
```

Sample Output: Output is read as in Table 8.7

Table 8.7: Movie feature pairs

$movie_i$	$rating_i$	$numRatings_i$	$movie_j$	$rating_j$	$numRatings_j$
1	5.0	392	2	3.0	121
1	5.0	392	3	4.0	85
1	5.0	392	4	3.0	198
1	5.0	392	5	3.0	79
1	5.0	392	6	5.0	23
1	5.0	392	7	4.0	346
1	5.0	392	8	1.0	194
1	5.0	392	9	5.0	268
1	5.0	392	10	3.0	82
1	5.0	392	11	2.0	217

Step 6: With reference to the Correlation function:

$$Corr(X,Y) = \frac{n\sum xy - \sum x \sum y}{\sqrt{n\sum x^2 - (\sum x)^2}\sqrt{n\sum y^2 - (\sum y)^2}}$$

First, We need to calculate the $X \cdot Y$, X^2 and Y^2.

```
import com.twitter.scalding._
class RecommendMovies(args : Args) extends Job(args) {
val ratings = Tsv("ua.base")
  .read
  .mapTo((0, 1, 2) -> ('user, 'movie, 'rating)) {
    fields : (Int, Int, Double) => fields
  }

val ratingsWithSize = ratings
  .groupBy('movie) { _.size('numRaters) }
  .write(Tsv("ratingsWithSize.tsv"))

val ratingsJoinWithSize = ratings
  .rename('movie -> 'movieX)
  .joinWithLarger('movieX -> 'movie, ratings)
  .discard('movieX)


val ratings2 = ratingsJoinWithSize
  .rename(('user, 'movie, 'rating, 'numRaters) ->
    ('user2, 'movie2, 'rating2, 'numRaters2))

val ratingPairs = ratingsJoinWithSize
  .joinWithSmaller('user -> 'user2, ratings2)
  .filter('movie, 'movie2) {
    movies : (String, String) => movies._1 < movies._2}
  .project('movie, 'rating, 'numRaters, 'movie2,
             'rating2, 'numRaters2)
```

```
val vectorCalcs = ratingPairs
  .map(('rating, 'rating2) -> ('ratingProd, 'ratingSq,
    'rating2Sq)) {
    ratings : (Double, Double) =>
    (ratings._1 * ratings._2,
     scala.math.pow(ratings._1, 2),
     scala.math.pow(ratings._2, 2))
  }
  .write(Tsv("vectorCalcs.tsv"))
}
```

Second, we calculate:

- $\sum ratings_i \cdot ratings_j$

- $\sum ratings_i$

- $\sum ratings_j$

- $\sum ratings_i^2$

- $\sum ratings_j^2$

To implement this we need to groupBy the movie pairs. Scalding provides groupBy API that takes Fields as an argument and applies function on the grouped result.

```
import com.twitter.scalding._
class RecommendMovies(args : Args) extends Job(args) {
val ratings = Tsv("ua.base")
  .read
  .mapTo((0, 1, 2) -> ('user, 'movie, 'rating)) {
    fields : (Int, Int, Double) => fields
  }

val ratingsWithSize = ratings
  .groupBy('movie) { _.size('numRaters) }
  .write(Tsv("ratingsWithSize.tsv"))

val ratingsJoinWithSize = ratings
  .rename('movie -> 'movieX)
  .joinWithLarger('movieX -> 'movie, ratings)
  .discard('movieX)

val ratings2 = ratingsJoinWithSize
  .rename(('user, 'movie, 'rating, 'numRaters) ->
    ('user2, 'movie2, 'rating2, 'numRaters2))

val ratingPairs = ratingsJoinWithSize
  .joinWithSmaller('user -> 'user2, ratings2)
  .filter('movie, 'movie2) {
    movies : (String, String) => movies._1 < movies._2}
  .project('movie, 'rating, 'numRaters, 'movie2,
           'rating2, 'numRaters2)
```

```scala
val vectorCalcs = ratingPairs
  .map(('rating, 'rating2) -> ('ratingProd, 'ratingSq,
    'rating2Sq)) {
    ratings : (Double, Double) =>
    (ratings._1 * ratings._2,
     scala.math.pow(ratings._1, 2),
     scala.math.pow(ratings._2, 2))
  }
  .groupBy('movie, 'movie2) {
    _.spillThreshold(500000)
    .size // length of each vector
    .sum[Double]('ratingProd -> 'dotProduct)
    .sum[Double]('rating -> 'ratingSum)
    .sum[Double]('rating2 -> 'rating2Sum)
    .sum[Double]('ratingSq -> 'ratingNormSq)
    .sum[Double]('rating2Sq -> 'rating2NormSq)
    .max('numRaters)
    .max('numRaters2)
    }
  .write(Tsv("vectorCalcs.tsv"))
}
```

Important: Note the spillThreshold option in the groupBy construct, this allows us to set the number of Keys and overrides the default Threshold value set for the AggregateBy function. This allows for number of keys that can be stored in the memory before calculation.

Step 7: We implement the Correlation expression as a function in Scala.

```scala
def correlation(size : Double,
                dotProduct : Double,
                ratingSum : Double,
                rating2Sum : Double,
                ratingNormSq : Double,
                rating2NormSq : Double) = {

    val numerator = size*dotProduct - ratingSum*rating2Sum
    val denominator =
    scala.math.sqrt(size*ratingNormSq - ratingSum*ratingSum) *
    scala.math.sqrt(size*rating2NormSq - rating2Sum*rating2Sum)

    numerator / denominator
    }
```

Applying the Correlation function on the resultant Fields would explain the dependance between the two movie vectors.

```scala
import com.twitter.scalding._
class RecommendMovies(args : Args) extends Job(args) {
val ratings = Tsv("ua.base")
  .read
  .mapTo((0, 1, 2) -> ('user, 'movie, 'rating)) {
    fields : (Int, Int, Double) => fields
  }

val ratingsWithSize = ratings
  .groupBy('movie) { _.size('numRaters) }
  .write(Tsv("ratingsWithSize.tsv"))

val ratingsJoinWithSize = ratings
  .rename('movie -> 'movieX)
  .joinWithLarger('movieX -> 'movie, ratings)
  .discard('movieX)


val ratings2 = ratingsJoinWithSize
  .rename(('user, 'movie, 'rating, 'numRaters) ->
    ('user2, 'movie2, 'rating2, 'numRaters2))

val ratingPairs = ratingsJoinWithSize
  .joinWithSmaller('user -> 'user2, ratings2)
  .filter('movie, 'movie2) {
     movies : (String, String) => movies._1 < movies._2}
  .project('movie, 'rating, 'numRaters, 'movie2,
              'rating2, 'numRaters2)

val vectorCalcs = ratingPairs
  .map(('rating, 'rating2) -> ('ratingProd, 'ratingSq,
    'rating2Sq)) {
    ratings : (Double, Double) =>
    (ratings._1 * ratings._2,
     scala.math.pow(ratings._1, 2),
     scala.math.pow(ratings._2, 2))
  }
  .groupBy('movie, 'movie2) {
    _.spillThreshold(500000)
    .size // length of each vector
    .sum[Double]('ratingProd -> 'dotProduct)
    .sum[Double]('rating -> 'ratingSum)
    .sum[Double]('rating2 -> 'rating2Sum)
    .sum[Double]('ratingSq -> 'ratingNormSq)
    .sum[Double]('rating2Sq -> 'rating2NormSq)
    .max('numRaters)
    .max('numRaters2)
      }
val similarities = vectorCalcs
    .map(('size, 'dotProduct, 'ratingSum, 'rating2Sum,
          'ratingNormSq, 'rating2NormSq, 'numRaters,
          'numRaters2) ->
          ('correlation)) {
```

```
        fields : (Double, Double, Double, Double, Double,
                  Double, Double, Double) =>

      val (size, dotProduct, ratingSum, rating2Sum,
           ratingNormSq, rating2NormSq, numRaters,
           numRaters2) = fields

      val corr = correlation(size, dotProduct, ratingSum,
             rating2Sum, ratingNormSq, rating2NormSq)

      (corr)
      }
      .write(Tsv("similarties.tsv"))
def correlation(size : Double,
                dotProduct : Double,
                ratingSum : Double,
                rating2Sum : Double,
                ratingNormSq : Double,
                rating2NormSq : Double) = {

    val numerator = size*dotProduct - ratingSum*rating2Sum
    val denominator =
    scala.math.sqrt(size*ratingNormSq - ratingSum*ratingSum) *
    scala.math.sqrt(size*rating2NormSq - rating2Sum*rating2Sum)
    numerator / denominator
  }

}
```

Step 8: Apart from the general Correlation similarity there are other similarity functions that can used to gain further insight in to the dependence between the movie vectors refer Section 8.2.1:

- Regularized Correlation
- Cosine Similarity
- Jaccard Similarity

```
import com.twitter.scalding._
class RecommendMovies(args : Args) extends Job(args) {
val ratings = Tsv("ua.base")
  .read
  .mapTo((0, 1, 2) -> ('user, 'movie, 'rating)) {
    fields : (Int, Int, Double) => fields
  }

val ratingsWithSize = ratings
  .groupBy('movie) { _.size('numRaters) }
  .write(Tsv("ratingsWithSize.tsv"))

val ratingsJoinWithSize = ratings
  .rename('movie -> 'movieX)
```

```scala
  .joinWithLarger('movieX -> 'movie, ratings)
  .discard('movieX)


val ratings2 = ratingsJoinWithSize
  .rename(('user, 'movie, 'rating, 'numRaters) ->
    ('user2, 'movie2, 'rating2, 'numRaters2))

val ratingPairs = ratingsJoinWithSize
  .joinWithSmaller('user -> 'user2, ratings2)
  .filter('movie, 'movie2) {
     movies : (String, String) => movies._1 < movies._2}
  .project('movie, 'rating, 'numRaters, 'movie2,
              'rating2, 'numRaters2)

val vectorCalcs = ratingPairs
  .map(('rating, 'rating2) -> ('ratingProd, 'ratingSq,
    'rating2Sq)) {
     ratings : (Double, Double) =>
     (ratings._1 * ratings._2,
      scala.math.pow(ratings._1, 2),
      scala.math.pow(ratings._2, 2))
  }
  .groupBy('movie, 'movie2) {
    _.spillThreshold(500000)
    .size // length of each vector
    .sum[Double]('ratingProd -> 'dotProduct)
    .sum[Double]('rating -> 'ratingSum)
    .sum[Double]('rating2 -> 'rating2Sum)
    .sum[Double]('ratingSq -> 'ratingNormSq)
    .sum[Double]('rating2Sq -> 'rating2NormSq)
    .max('numRaters)
    .max('numRaters2)
      }

val PRIOR_COUNT = 10
val PRIOR_CORRELATION = 0

val similarities = vectorCalcs
  .map(('size, 'dotProduct, 'ratingSum, 'rating2Sum,
       'ratingNormSq, 'rating2NormSq, 'numRaters,
       'numRaters2) ->
         ('correlation, 'regularizedCorrelation,
          'cosineSimilarity, 'jaccardSimilarity)) {

  fields : (Double, Double, Double, Double,
            Double, Double, Double, Double) =>

  val (size, dotProduct, ratingSum, rating2Sum,
       ratingNormSq, rating2NormSq, numRaters,
       numRaters2) = fields

  val corr = correlation(size, dotProduct,
    ratingSum, rating2Sum, ratingNormSq, rating2NormSq)
```

```scala
  val regCorr = regularizedCorrelation(size, dotProduct,
            ratingSum, rating2Sum, ratingNormSq,
            rating2NormSq, PRIOR_COUNT, PRIOR_CORRELATION)

  val cosSim = cosineSimilarity(dotProduct,
                    scala.math.sqrt(ratingNormSq),
                    scala.math.sqrt(rating2NormSq))

  val jaccard = jaccardSimilarity(size, numRaters,
                                       numRaters2)

  (corr, regCorr, cosSim, jaccard)
}
 .write(Tsv("similarties.tsv"))

def correlation(size : Double,
                dotProduct : Double,
                ratingSum : Double,
                rating2Sum : Double,
                ratingNormSq : Double,
                rating2NormSq : Double) = {

  val numerator = size*dotProduct - ratingSum*rating2Sum
  val denominator =
  scala.math.sqrt(size*ratingNormSq - ratingSum*ratingSum) *
  scala.math.sqrt(size*rating2NormSq - rating2Sum*rating2Sum)
  numerator / denominator
  }

def regularizedCorrelation(size : Double,
                           dotProduct : Double,
                           ratingSum : Double,
                           rating2Sum : Double,
                           ratingNormSq : Double,
                           rating2NormSq : Double,
                           virtualCount : Double,
                           priorCorrelation : Double) = {

  val unregularizedCorrelation = correlation(size,
                    dotProduct, ratingSum,
                    rating2Sum, ratingNormSq,
                    rating2NormSq)
  val w = size / (size + virtualCount)

  w*unregularizedCorrelation + (1-w)*priorCorrelation
}

def cosineSimilarity(dotProduct : Double,
                     ratingNorm : Double,
                     rating2Norm : Double) = {
  dotProduct / (ratingNorm * rating2Norm)
}
```

```scala
def jaccardSimilarity(commonRaters : Double,
                      raters1 : Double,
                      raters2 : Double) = {
  val union = raters1 + raters2 - commonRaters
  commonRaters / union
}


}
```

Problems

8.1. Download the Book-Crossing dataset [33]. Build a recommender system using the book ratings in the dataset using Spark and Scalding. Use the Spark Programming guide explained in chapter 4.

References

1. Resnick, P., Varian, H.R.: *Recommender systems.* Communications of the ACM 40(3), 56-58 (1997)
2. Burke, R.: *Hybrid web recommender systems. In: The Adaptive Web*, pp. 377-408. Springer Berlin / Heidelberg (2007)
3. Jannach, D.: *Finding preferred query relaxations in content-based recommenders.* In: 3rd International IEEE Conference on Intelligent Systems, pp. 355-360 (2006)
4. Mahmood, T., Ricci, F.: *Improving recommender systems with adaptive conversational strategies.* In: C. Cattuto, G. Ruffo, F. Menczer (eds.) Hypertext, pp. 73-82. ACM (2009)
5. McSherry, F., Mironov, I.: *Differentially private recommender systems: building privacy into the net.* In: KDD 09: Proceedings of the 15th ACM SIGKDD international conference on Knowledge discovery and data mining, pp. 627-636. ACM, New York, NY, USA (2009)
6. Schwartz, B.: *The Paradox of Choice.* ECCO, New York (2004)
7. Ricci, F.: Travel recommender systems. IEEE Intelligent Systems 17(6), 55-57 (2002)
8. Herlocker, J., Konstan, J., Riedl, J.: Explaining collaborative filtering recommendations. In: In proceedings of ACM 2000 Conference on Computer Supported Cooperative Work, pp. 241-250 (2000)
9. Brusilovsky, Peter. *Methods and techniques of adaptive hypermedia.* User modeling and user-adapted interaction 6.2-3 (1996): 87-129.
10. Montaner, M., Lopez, B., de la Rosa, J.L.: *A taxonomy of recommender agents on the Internet.* Artificial Intelligence Review 19(4), 285-330 (2003)
11. Fisher, G.: *User modeling in human-computer interaction.* User Modeling and User-Adapted Interaction 11, 65-86 (2001)
12. Berkovsky, S., Kuflik, T., Ricci, F.: *Mediation of user models for enhanced personalization in recommender systems.* User Modeling and User-Adapted Interaction 18(3), 245-286 (2008)
13. Taghipour, N., Kardan, A., Ghidary, S.S.: *Usage-based web recommendations: a reinforcement learning approach.* In: Proceedings of the 2007 ACM Conference on Recommender Systems, RecSys 2007, Minneapolis, MN, USA, October 19-20, 2007, pp. 113-120 (2007)

14. Schafer, J.B., Frankowski, D., Herlocker, J., Sen, S.: *Collaborative filtering recommender systems*. In: The Adaptive Web, pp. 291-324. Springer Berlin / Heidelberg (2007)
15. Adomavicius, G., Sankaranarayanan, R., Sen, S., Tuzhilin, A.: I*ncorporating contextual information in recommender systems using a multidimensional approach*. ACM Trans. Inf. Syst. 23 (1), 103-145 (2005)
16. Mitchell, T.: *Machine Learning*. McGraw-Hill, New York (1997)
17. Konstan, J.A., Miller, B.N., Maltz, D., Herlocker, J.L., Gordon, L.R., Riedl, J.: *GroupLens: applying collaborative filtering to usenet news*. Communications of the ACM 40 (3), 77-87 (1997)
18. Linden, G., Smith, B., York, J.: Amazon.com recommendations: Item-to-item collaborative filtering. IEEE Internet Computing 7 (1), 76-80 (2003)
19. Breese, J.S., Heckerman, D., Kadie, C.: Empirical analysis of predictive algorithms for collaborative filtering. In: Proc. of the 14th Annual Conf. on Uncertainty in Artificial Intelligence, pp. 43-52. Morgan Kaufmann (1998)
20. Hofmann, T.: Collaborative filtering via Gaussian probabilistic latent semantic analysis. In: SIGIR 03: Proc. of the 26th Annual Int. ACM SIGIR Conf. on Research and Development in Information Retrieval, pp. 259-266. ACM, New York, NY, USA (2003)
21. Zitnick, C.L., Kanade, T.: Maximum entropy for collaborative filtering. In: AUAI 04: Proc. of the 20th Conf. on Uncertainty in Artificial Intelligence, pp. 636-643. AUAI Press, Arlington, Virginia, United States (2004)
22. Pazzani, M.J.: A framework for collaborative, content-based and demographic filtering. Artificial Intelligence Review 13 (5-6), 393-408 (1999)
23. Bridge, D., Goker, M., McGinty, L., Smyth, B.: Case-based recommender systems. The Knowledge Engineering review 20 (3), 315-320 (2006)
24. Sinha, R.R., Swearingen, K.: Comparing recommendations made by online systems and friends. In: DELOS Workshop: Personalisation and Recommender Systems in Digital Libraries (2001)
25. Groh, G., Ehmig, C.: Recommendations in taste related domains: collaborative filtering vs. social filtering. In: GROUP 07: Proceedings of the 2007 international ACM conference on Supporting group work, pp. 127-136. ACM, New York, NY, USA (2007)
26. Sarwar, B., Karypis, G., Konstan, J., Riedl, J.: Incremental singular value decomposition algorithms for highly scalable recommender systems. In: Proceedings of the 5th International Conference in Computers and Information Technology (2002)
27. Ramakrishnan, N., Keller, B.J., Mirza, B.J., Grama, A., Karypis, G.: When being weak is brave: Privacy in recommender systems. IEEE Internet Computing cs.CG/0105028 (2001)
28. Herlocker, J., Konstan, J., Borchers, A., Riedl, J.. *An Algorithmic Framework for Performing Collaborative Filtering*. Proceedings of the 1999 Conference on Research and Development in Information Retrieval. Aug. 1999.
29. *Cascading Pipes* http://docs.cascading.org/cascading/1.2/javadoc/cascading/pipe/Pipe.html
30. Singhal, Amit. *Modern Information Retrieval: A Brief Overview*. Bulletin of the IEEE Computer Society Technical Committee on Data Engineering 24 (4): 35-43, 2001
31. Tan, Pang-Ning; Steinbach, Michael; Kumar, Vipin, *Introduction to Data Mining*, ISBN 0-321-32136-7, 2001
32. Yule, G.U and Kendall, M.G., *An Introduction to the Theory of Statistics*, 14th Edition (5th Impression 1968). Charles Griffin & Co. pp 258-270, 1950
33. Cai-Nicolas Ziegler, Book-Crossing dataset, [Online] Available: http://www2.informatik.uni-freiburg.de/ cziegler/BX/

Index

© Springer International Publishing Switzerland 2015
K.G. Srinivasa and A.K. Muppalla, *Guide to High Performance Distributed Computing*, Computer Communications and Networks, DOI 10.1007/978-3-319-13497-0

Printed in the United States
By Bookmasters